Leonard Williams was born in London in 1910. Into an exciting life he has packed many professions, notably music, zoo-keeping and primate ethology. As a distinguished jazz and classical musician (father and teacher of the guitarist John Williams) he worked for many years with the Australian Broadcasting Commission and he was founder of the Spanish Guitar Centre in London. He now lives with his family and a colony of woolly monkeys at the Monkey Sanctuary near Looe in Cornwall.

BY THE SAME AUTHOR
The Dancing Chimpanzee
Challenge to Survival

Leonard Williams

SAMBA

ALLISON & BUSBY
London and New York

Samba and the Monkey Mind revised edition
published 1980 by
Allison & Busby Ltd.,
6a Noel Street, London W1V 3RB., England.,
and distributed in the USA by
Schocken Books Inc.,
200 Madison Avenue, New York, N.Y. 10016.

British Library Cataloguing in Publication Data

Williams, Leonard
  Samba and the Monkey Mind

  1. Woolly monkeys
  I. Title
599'.82   QL737.P925   80-40597

ISBN 0-85031-339-2
ISBN 0-85031-340-6 Pbk

Printed in Great Britain by A. Wheaton & Co. Ltd., Exeter

Contents

PHOTOGRAPHS
from the author's collection

AUTHOR'S NOTE

In English-speaking countries, *Lagothrix Lagotricha* (Humboldt) are popularly known as "woolly monkeys".

Although there are many types, each with distinct physical characteristics and varying in coloration from grey to brown, my own references are specifically to the grey types. The term "woolly monkey" is retained throughout this book as being more familiar to English-speaking readers.

I am especially indebted to Dr Konrad Lorenz, Lord Clark and Sir Julian Huxley for their valuable suggestions in the preparation of this book.

Passages from *King Solomon's Ring* by Konrad Lorenz and *My Friends the Baboons* by Eugène Marais are reproduced by permission of Methuen and Co. Ltd.; from *Ring of Bright Water* by Gavin Maxwell by permission of Longmans, Green.

Murrayton, Cornwall   LEONARD WILLIAMS

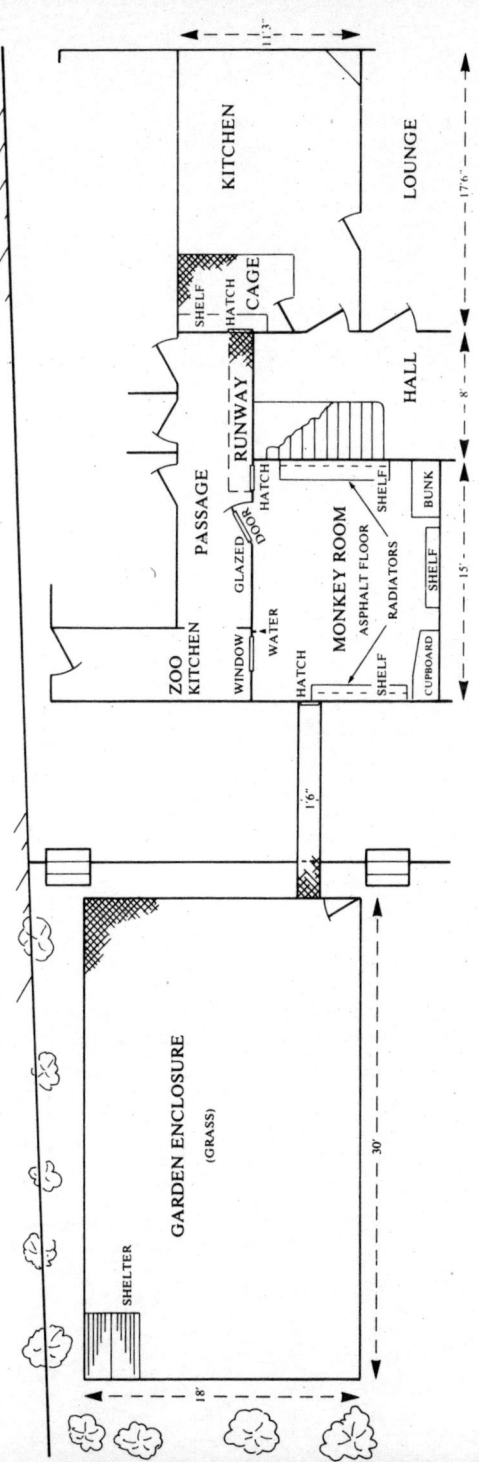

PLAN SHOWING MONKEY SECTION OF HOUSE

PART 1

The Woolly Monkey

1

Green Fingers

If a neighbour were to ask: "Who was the young man I saw in your garden yesterday?" we are not likely to reply: "A Nordic European of the genus *Homo sapiens*. He is bimanous and bipedal, and has opposite thumbs." But since we have friends that induce some people to enquire: "What are those black things out there in your garden?" an explanation of some kind must be given. We say they are the primates, *Lagothrix lagotricha*, of the group Ceboids. But should someone ask: "What's the name of those lovely little black monkeys?" we reply: "Woolly grey monkeys, from the Amazon jungle in South America."

When people ask me why I keep monkeys, I never quite know what to say. The question embarrasses me and makes me feel that I am perhaps eccentric. But I do know when my interest in woolly monkeys began. As a young man I developed the somewhat unusual habit of visiting the zoo, not once a year on Bank Holiday Mondays, but almost every weekend throughout the year. I cannot say what gave rise to this habit, unless it was the simple fact that the happiest and most exciting days of my childhood were those spent at the zoo. It is very easy to speak of turning points in our lives, of people and events that shook our thinking or changed our hearts, without realising that an accumulation of previous events and influences had already prepared us for such

changes. So true is this, we are often attracted by some things
without knowing or even caring why, and such was my state
of mind during those early days at the zoo. I usually went
alone, for whenever I persuaded a friend to go with me my
habit of remaining by one cage for a considerable time would
drive my companion to despair.

There was hardly an animal at the zoo I did not know,
and it was a coincidence that I happened to meet "him" on
the very day of his arrival. He had no name, unless you call
Lagothrix lagotricha a name. The keeper could give me no
information, and seemed surprised by my interest. "It" was
just another monkey in the monkey-house. But he was
nothing of the sort; he was a woolly monkey from the
Amazon jungle, and the most beautiful monkey I had ever
seen. His eyes were magnificently expressive, and he was
sad, and he sobbed, as woollies do when you whisper to
them.

There is a special beauty about the woolly monkeys that
is hard to name. In repose they are contemplative, serene and
plaintive, as though Nature has chosen them to look sad on
her account. In many ways they are more human-like than
other monkeys.

Woolly monkeys are found in most parts of tropical
America, ranging from southern Venezuela, through Brazil,
Ecuador and Peru. The main feature that distinguishes them
from the Old World monkeys is the prehensile tail. This tail
has a pad-like underside at its end, from six to seven inches
long, the texture of which is not unlike that of kid leather. It is
a most beautiful and remarkable organ, and is often referred
to as a hand-tail. Woolly monkeys can hang by this tail alone,
and for long periods. They also use it for picking up, carrying
and reaching out for objects that are beyond the range of their

hands, in much the same way as an elephant uses his trunk. Almost everything the elephant can do with his trunk the woolly monkey can manage with his tail. Even with an object as difficult to hold as a ball, the end of this sensitive tail somehow manages to curl itself round it and grasp it as firmly and securely as any hand; and when the owner moves along, up goes the tail in the air as though it has a life of its own and was resigned to going wherever the rest of the monkey chose to take it.

When you have lived for a long time with woolly monkeys you feel such respect for the prehensile tail that you begin to regard it as a separate personality. Sometimes Samba will climb on to my lap and lean back, lazily and contentedly, her arms and legs sprawled out, as she sinks dreamily into a pensive mood. There she sits, utterly relaxed, her arms outstretched and her back resting against me, peering across her fat belly at her feet and—that tail! Is *he* pensive? Not a bit! He is about to start a merry little game of his own. He rolls up his prehensile tip into a neat little curl and makes a sideways jump and back again. Samba watches with mild interest. He draws himself up, so slowly, moves a little from side to side, and then, quite suddenly, flops down on to my hand. He, too, has another side to his nature, and after a hard day's work in the garden, swinging, hanging, twisting and turning, he wants to enjoy an idle moment no less than Samba. Shall we stroke him, or tease him? We move our hand away and allow it to fall on top of him, tit for tat. He follows suit immediately. Now he moves up again and over to Samba, hovering in front of her face. Suddenly, she grabs him with both hands as though catching a fly, and laughs in the way that woollies do, with a "huh, huh, huh" and a great shaking of the head. A dog will play with his tail, but here it seems the tail plays with the monkey. There are very few of Samba's moods in which her tail does not play an important part.

The strength of the prehensile tail is quite staggering. If the fully-grown male in a fit of rage throws it round your neck you will be savaged almost to death before you can be free of it. Attached to this tail is a body of incalculable strength that can weigh as much as thirty pounds, with four long and sinewy limbs to use against you. Each limb has a devil's hand on the end of it, and while two of these hands dig and tear at your eyes and face, the other two and the tail will fasten on to your arms so that you can barely move them in order to defend yourself. Added to all this fighting machinery is a mouth more dangerous than that of the lynx; not because of the long canines that protrude over its lower jaw and their interlocking action when they bite and drag, but because of the cunning of the primate brain that drives them into your skull, your face or your throat. This brain knows what it is doing, and those powerful fingers are gouging into your eyes because a blind opponent is easier to kill.

In the early days we had three monkeys: Liz, Samba, and Lulu. No one in this country had been successful in establishing a breeding pair, though there had been two woolly monkeys stillborn at Molly Badham's zoo in Staffordshire. We wanted to discover whether the special environment we had built for our monkeys would be suitable for mating, and Molly Badham was kind enough to let us have one of her mature males for this purpose. That is how Pepi, at the age of six, came to join our woolly monkey community as a husband for Lulu, the most mature of the females.

I therefore set out to achieve a special relationship with Pepi, one that would allow me to continue my friendship with

Lulu in the event of her having a baby. All went well for the first year, during which time Pepi had married Lulu and made it clear that she belonged to him and no one else. The honeymoon went on for months, it fact it never stopped. In that year, like most married men, Pepi put on weight, but it was the weight of more sinew and muscle and of the development from adolescence to manhood. His shoulders broadened, his forearms thickened, his canines grew, and at the age of seven, when he weighed twenty-five pounds, the mere sight of him was enough to raise a doubt in my mind. But we continued to be friends and everything pointed to the fact that he still trusted and respected me.

When I take my mind back to that terrible day, I often wonder whether, had I not turned my back on Pepi and knelt down to pick up a bucket from the floor, the attack would have been made. I remember that, as I turned, he was lying on a high shelf by the window, looking quite relaxed, and that Lulu was in the outside enclosure. Then it happened: no warning, no sound, just a heavy weight that fell on my shoulders, a great tail that went round my neck, two long legs pinning my sides, two powerful arms wrapped round my head and face, and a strange pressure on my skull.

When you are standing and an animal attacks from the front you stand a chance, even a good chance if you keep your wits about you. But what can you do when a twenty-five-pound death machine wraps itself round your neck and your shoulders, and there you are, crouching on the ground, with a bucket in one hand and a cigarette in the other? I did nothing, absolutely nothing, while I counted, very slowly up to ten. Pepi also did nothing. You can do a lot of thinking in ten seconds, and besides trembling with fear and working out the best plan of counter-attack and escape I was thinking about such niceties as main arteries, the jugular vein, and damaged eyes. I also thought about my wife and children, and those docile, affectionate little friends of ours—the little woolly people!

Ten seconds are up, so I talk to Pepi. "Hello, old man. There's a good Pepi. Huh, huh, huh! How about a game?" He does nothing. How long do I have to remain sitting here, just waiting—five minutes, ten? (As a zoo-keeper, I had been in trouble before, trapped in a hippo den. There was one way in, and daylight went out as two ton of hippopotamus appeared in the entrance. That was another time when I counted up to ten, and said my prayers as well, only then the hippo came right in, looked straight at me, turned round and walked out again.) Still, Pepi has done nothing serious as yet, and I can hardly sit here all day, so let's try standing up very slowly. Not bad, a little more pressure from the tail, but nothing to worry about. Now for a different tactic—be masterful! "Off, Pepi!" Pepi tightens his hold and champs his jaws.

And then my nerve broke. I lifted my arms with the object of beating him off, when he seemed to leap away from me. What a stroke of luck! I made for the door, but he came at me across the ground. There was not enough time to open the door, so I turned and kicked him, full in the face. It was a foolish thing to do, because even in play woolly monkeys will always grasp whatever is thrust at them: feet, hands, brooms, anything. They just hang on, and it is extremely difficult to shake them off. But I was lucky, for the kick connected, and though it made little impression on him, it gave me the few seconds I needed to get out of the room and lock the door.

I was in no pain, and I began to congratulate myself on a miraculous escape, when I saw my hand. I could not believe my eyes, for it was torn across from near the centre of the palm. My eyes felt wet, and also my face and neck. Blood was pouring down from the top of my head, over my face and on to my clothes. So that was why Pepi had leapt away from me! With two highly successful drag bites, administered at lightning speed, he had decided to jump clear and prepare himself for the next stage of the fight. I was not only unaware of the fact that Pepi had bitten me, I had experienced no pain whatever.

No longer under the nervous strain of directing all my energy and alertness to coping with Pepi's attack, I was now able to take stock of what had really happened, and as I stood there, contemplating with horror the injury to my hand, a terrible pain began to creep into it. I will spare the reader the details of my recovery, and record simply that I was in fact an extremely fortunate man to escape with a scalp wound and a lacerated right hand.

By attempting a relationship with Pepi, I was defying the lessons of my own experiences and those of others who had attempted similar experiments. Here was a fully mature male, healthy and virile, not reared by me from babyhood, and so possessive about Lulu that even the wild birds in the garden were not allowed to hop too close to her. Yet I would enter inside the monkey-room, with Lulu there, and turn my back on him, or I would go straight up to him while he was guarding Lulu, hold out my arms and say: "Poor old Pepi, I don't want your Lulu," and he would wrap his enormous tail round his own neck, bury his face in his arms, and shake with emotion while I consoled him, with Lulu still sitting there only a few feet away. I knew that I was challenging a million years of instinct with an affection acquired in a single life-time, but I believed that the attack, if it came at all, would occur in precisely this situation, with Pepi clinging to me in his fear and seeking reassurance. Indeed I was convinced that if I continued to tolerate his emotional outbursts and made no advances to Lulu, he would be inhibited, disarmed and unlikely to attack. All I had to do was to stand there and console him, and in time he would learn to trust me.

In the light of Pepi's attack, my theory would seem to be proved false. Pepi, however, was an exceptionally aggressive male compared with other male monkeys I have known. In the chapter entitled "Character and Dominance" I have defined what I believe to be two different character-types of woolly monkey, and I am not yet convinced that my experiment would fail with a male of the Samba type—a physical type that is more intelligent and less aggressive. Considering that Pepi came to me at the age of six, a mature and semi-wild animal, his tolerance of my intrusion into his private life was indeed remarkable. I shall certainly try again with another male when we have built our new woolly monkey sanctuary in the country, and when we have found a suitable male for Samba. And next time I shall act with more caution and with greater understanding.

2
"Other Sheep Have I . . ."

For every baby monkey captured in the Amazon jungles a mother loses her life. The method of capture is to shoot the mother, and when she falls to the ground dead take the baby, if still alive, from her back. Torn from a natural environment where they would remain with their mothers and be nourished by them for as long as fifteen months, these four-month-old babies are packed in small crates and sent on a journey of thousands of miles. Many die on the way, and those who do survive arrive at the airport half dead or in a very poor state of health.

Almost every case known to me of private owners who have endeavoured to rear these baby monkeys has followed the same tragic pattern, a struggle of life and death; yet they are the only ones who stand a reasonable chance of regaining health and reaching adulthood. The mature and healthy woolly monkey that one occasionally sees at the zoo has nearly always been presented by someone who reared it with all the comforts of a home and garden and on a superior and more comprehensive diet.

The importance of a substitute mother for a baby woolly monkey is made clear in a letter we received from a Mrs Rex. The extract here explains how she came by her woolly monkey Samba, the very same Samba who belongs to us today for Mrs Rex died two years ago and left Samba to our care: "I received a letter from an animal importer to say he had a tiny baby woolly for sale, so I went to see it . . . There in a huge cage, crouched in the corner in the sawdust, with its tail covering its tiny face, was the smallest monkey I have ever seen, looking cold and sad, no hair on her tail, and very little elsewhere. I lifted her out and she looked at me and said a very weak 'eelk' and tried to crawl under my coat. I *had* to take her home with me . . . I made a little hammock and suspended it round my neck under my jumper (it was November) and there baby Samba lived and thrived, for she was too young to leave her mother. Day by day she gained a little, her hair grew, and she got cheeky with my dogs, and at long last she climbed a rope up to the ceiling!"

So Samba lived, and today at the age of five she is one of the loveliest female woolly monkeys you ever saw. She and Liz are the greatest of friends, and will end their days here; both indeed may well outlive the author.

Woolly monkeys are not so delicate as some "authorities" would have us believe. They are highly emotional and responsive and become sick and dejected under average zoo conditions. With fresh air, plenty of space, swings, ropes, play material, and free access at all times to warm inside quarters, they will thrive under the most severe climatic conditions and changes.

When young monkeys are healthy, they are extremely tough. Little Jimmy, eighteen months old and weighing only seven pounds, once fell from the top beam of the outside cage, flat on his middle. It was his own fault, for he had been trying to outwit Liz, who is older and stronger. Overcome with the satisfaction of getting home one good nip and making Liz squeal, Jimmy had taken a flying leap at her while she was hanging by the tail from the top beam. He missed, which is something I would not have thought possible for a woolly monkey, for whatever else he is doing, especially in high places, that tail is always there, anchored to something. We had sown new grass seed in the garden enclosure, and Jimmy fell thirteen feet on to hard earth. He lay on the ground, hardly moving, and croaked for several minutes. By the time I had reached him he was on all fours, with a strong list to starboard and his head twisted to one side. He crawled along like a broken insect. Then he stopped and was violently sick. I was so distressed and sorry for him that it was only by the greatest effort of will that I could leave him unattended for a while, this being the best way of observing the extent of his injuries. Quite suddenly he sat down and began to look more comfortable. A second later, he started to climb the pole, slowly at first, then faster as he neared the top. As Liz approached he shook his head and sobbed, and they huddled close together. The huddle developed into "huh, huh, huh" and then into "aaarrk", and the game was on again.

A little while later, when the game was over and resolved, Jimmy settled down to a meal of rice pudding, half a lettuce, and two pieces of malt-bread and butter. He then looked up at me, and said "eelk", which is woolly monkey language for "Hello! I'm feeling fine", and I thought of Charlie Chaplin, who, above all artists, has shown us that we need more than the will to live, we need the will to be cheerful about it—come what may.

Our woolly monkeys—Liz, Samba, Lulu, and Jimmy— all reached adulthood, and today they enjoy perfect health. Liz and Lulu both had enteritis when they were eighteen months old, and both recovered without the use of antibiotics. Veterinary officers in zoos consider enteritis to be generally fatal for woolly monkeys; even with antibiotic treatment, recovery is rare.

What chance would a human infant have of surviving enteritis if it were left alone in a cage to care for itself, with no attention of any kind? I have seen baby woolly monkeys in the dealers' shops, huddled up in corners, covered with cage filth, cold, sick and close to death. I have bent down close to the cage, whispering "eelk" and "huh, huh", and watched their faces light up as they answered with a tiny "eelk" and came crawling over towards me. What a great day of rejoicing it would be if these little babies were to find their way back to their mothers!

When Lulu was sick with enteritis she rested in a portable cot which we put close to a heater in our kitchen. We tempted her with delicacies and allowed her to have anything she wanted. Within a few days she took a little orange juice, a few grapes and a small piece of bread and butter. In two weeks she was well enough to join the monkey community. Liz was also taken sick and recovered under the same treatment.

In captivity woolly monkeys are not very intelligent about what they eat. Given the opportunity they are likely to eat poisonous plants native to this country. They are mainly vegetarians and are fond of almost every kind of fruit and green leaf—cabbage, lettuce, dandelion, dock and willow herb— and also beans and lentils. The only animal proteins we give to our monkeys are four ounces of boiled chicken and two eggs per week for each monkey. Roughage in the form of twigs, rope fibre and hay must be available at all times.

Until successful breeding lines have been established in this country, I am in principle against having the woolly monkey as a pet; but if you are unable to wait until that happy day arrives, then at least make sure you get a female, and remember that you will have to give it almost as much care as you would a human baby, just as Mrs Rex did with Samba.

I would like to end this chapter with a tribute to Mrs Rex. It is sad to think that she died without knowing that in her many letters to us she had written a very important piece of woolly monkey history. Perhaps one day I shall be given the opportunity of publishing these letters, for they represent a treasure trove of old-world reflections on the wild life that surrounded her little house in Weeley Heath, Essex, where she lived with her many animal friends, her flowers, her books, her memories and the spectacles that Samba was always breaking. The extracts that follow are taken from her last letter to us:

"When Samba was still quite a baby, Tail seemed to be a separate entity. Often she would sit with a toy in her hand, and Tail would creep over her shoulder and snatch the toy, much to Samba's dismay! It was a long time before she was able to relate cause to effect.

It is wonderful to come home from shopping and hear a little voice say 'eelk'. When I open the door two little arms go round my neck and a sweet little black suede face presses against mine. With squeaks and sobs and a great shaking of the head, she says: 'What a long, long time you have been away!' And then the dogs greet me, and Raffles the parrot says 'Allo, allo', and I am home. Samba is such a dear, and sometimes I think that when Christ said: 'Other sheep have I that are not of this flock', he meant these funny little woolly people, who cheer us so much.

The grass was so long this spring, my mower

would not cut it, so I answered an advertisement in the local paper for a woman gardener. She made a very good job of it, and when I took out a tray of tea for her, Samba came trotting along with me. Samba didn't like the sound of the motor-scythe, and was completely terrified of *her*. She had a very red and shiny face, a hat with a pheasant's feather, which wobbled all the time, a very short checked and patched coat, and under the coat a pair of enormous buttocks—looking just like the oldest inhabitant's prize marrows at the chapel sale. She also had shorts, men's socks and black boots. In a deep, deep voice, she turned to Samba and said: 'Allo'. Samba looked at her with eyes popping out, and with a shrill scream rushed up my skirt and vanished under my jumper. She was really stiff with terror, and I was at a loss to know how to get her over it.

Every time I now start my own mower I can hear Samba scream in her room, and when I go to console her, there she is, hunched in a corner, shaking and shivering, and it takes all day to wear off. Even then we have to go all round the garden, into every shed, upstairs and under the beds, in the cupboards and wardrobes, before Samba will believe *she* is not here! And if I cut the grass in the evening when Samba is asleep, her first thought in the morning is to rush to each window and glare into the garden in search of *her*.

Today the crocus and violets were out wide to the sun, and even a few daffodils, so I went for a walk in the lane with Samba and the dogs, and met the vicar of a neighbouring parish. He stopped me and said:

'I don't know how you can bear to have that *thing* crawling on you.' I was much taken aback, for I am a quiet person and inclined to peace, but I said: 'Vicar, the same God who made you made Samba as well, and I am sure He loves us all.' My own vicar loves Samba, and when he leaves after one of his parochial visits, I am never quite sure to whom the visit was paid!"

3

Eelk!

"Where's your teddy, Samba?" Samba replies with a joyful, high-pitched "eelk". She goes to the cupboard, opens the door and rummages through a collection of playthings belonging to the children until teddy is found. He is then hoisted on to her shoulders and held there while she walks round the room in an upright position. Or she will bring teddy to you and curl up with him in your lap, making little sobbing sounds and chattering with her teeth as though overcome with emotion.

"Eelk" and the sobbing and teeth-chattering ritual are allied with many moods, and have roots that lie deep in the sexual instincts. With a human being an infinite variety of moods and emotions is given a distinctive and expressive form, but with a woolly monkey one mood will often merge with another, leading to a whole sequence of moods. The end may be in domination, submission, reconciliation, or sexual consummation; or it can stop short of any one of these.

Let us say that Liz is in trouble with Samba, who incidentally, is the boss. Liz has probably made the great mistake of helping herself to some grapes without Samba's permission. Samba has crept up behind and delivered a small but decidedly unpleasant nip on Liz's rear. Liz screams with rage, but Samba nonchalantly turns her back and proceeds to eat the grapes as though Liz were not there. Liz is now inconsolable. She has been bitten and she is not going to get any grapes. When I appear on the scene she jumps on my back, curls herself round my neck and sobs, with the forearm shielding her eyes, manifesting all the same signs of affection for me as Samba does for her teddy. I now approach Samba, who is well aware of my intentions and tries to make off with the grapes. I grab her by the tail and give her a few smart whacks, whereupon she screams, drops the grapes, and collapses into my arms in the sobbing ritual. Liz, during Samba's punishment, has become almost delirious with satisfaction, evidenced by the woolly monkey's special war-cry, "nyonk, nyonk, nyonk". But with Samba sobbing her penitence quickly follows suit, and I am left standing there, embraced by two grief-stricken females.

Life continues, but peace does not stay for long. Liz and Samba are now becoming sensual, and if I allow this to continue we shall be back where we started, with another war on our hands. As the dominant male in this drama, I am expected to choose between Liz and Samba, and the only way out of this dilemma is for me to produce more grapes. Liz and Samba are delighted with the change in the programme and respond with a chorus of "eelks". I don't admit for one moment that this proves my sex-appeal is unable to compete with a grape. I think "eelk" in this context is monkey logic for "first things first", i.e. one can always make love, but one cannot always get grapes.

In an emergency woolly monkeys can usually be dispersed if one imitates the quiet cough-like grunt of the jaguar. This will send them running away and up to the highest place they can reach. But it is an unkind thing to do. They huddle together upon the topmost beams, their little black faces frozen in fear, making their danger call, "yook, yook, yook."

The richest part of monkey language is expressed not by sounds but by movements and gestures, as is so of most animals. When Samba has discovered a new toy and she is not quite sure whether it is alive or not, she will grasp it with the end of her tail, come running to me and jump on to my lap. There she will sit, waving the toy up and down with her tail for both of us to investigate, looking at me every so often to see what I am going to do. If I am uncooperative, she will drop the toy in my lap, and, if this fails, she will push my hand towards it. When I pick it up and show that it is safe to handle, she will then say "eelk" to indicate that she is satisfied with the reassurance. When I was in Australia I had a golden-crested cockatoo named Gerry, who would bring a ball to me by pushing it across the floor with his beak. When the ball was at my feet Gerry would stand there, swaying from side to side and inviting me to a game with a rhythmic thrusting back and forth of the head. Only when the game started did Gerry squawk with excitement and delight. Here we have a situation that is parallel with Samba and the toy horse. Both Gerry and Samba displayed a higher faculty of expression by movement and gesture than they could have achieved by sound. The sounds in monkey language that express elemental and instinctive feelings are nevertheless rich and varied in meaning, and it is interesting and important to interpret as many of them as we can.

The most common sound is unquestionably "eelk". In the most general terms it means "all is well". There are many kinds of "eelk", most of them sharp, high-pitched and joyful, with variations in pitch, inflexion, and strength, but they all stem from "all is well". For example, Liz is curled up in the top bunk, half asleep, and Samba is having a midnight snack on the shelf over by the window, for she likes to look at the stars while she eats her last orange for the night. Samba, for no apparent reason, unless it has just occurred to her that life is wonderful, says "eelk". Liz, whom we thought asleep, replies "eelk" so quickly as to sound like an echo. Translated this would mean:

Samba: "I'm feeling fine."

Liz: "So am I."

Or we might see Liz swinging by her tail in the far corner of the garden enclosure, by the apple tree. She hears Samba "eelk" some forty feet away, over the hatch which enters the monkey-room, and she answers with a loud and long "eeeooo . . . ." that begins as a crescendo and falls with a trill-like glissando into ". . . oolk". It is a most beautiful sound and in character and strength resembles the woolly monkey's call-cry. There can be little doubt that "eelk" is one of many signal sounds meaning "all is safe and well". "Eeeooolk" means that things are going particularly well.

One of the most intriguing sounds in woolly monkey language is "nyonk, nyonk, nyonk". It is carried on very much in the style of Morse code, and is an exceedingly difficult sound to interpret. In a free-for-all with "nyonks" coming from all directions, it often means somebody is getting the worst of it, and that what began as one against one, or two against two, has developed into three against one. But one cannot rely on this interpretation. If I stage a mock battle with my wife, June, and she pretends that I am hurting her, Samba will put a restraining hand on me, or sometimes in her confusion on June, adopting a protective attitude and "nyonk-ing" almost continuously. And when peace is restored Samba plays her part in the reconciliation with the sobbing ritual.

"Yook, yook" means danger, fear. If "eelk" is "yes",

then "yook" is "no". When woolly monkeys are frightened, whether by unfamiliar sounds in the night, shadows, reflections in the glass or moonlight, it requires only one to start for the others to join in. This "yooking" will often continue long after the cause of the disturbance has ceased. Fortunately for us this rarely happens, for when it does it is usually at night, and one of us has to get out of bed and go into the monkey-room to console and reassure them. This I do with a number of my own "eelks", which I am not very good at since I have difficulty in pitching my voice to anything like their register; but it usually works.

The call-cry of the woolly monkey must be one of the most beautiful sounds in the whole of nature. It is a sustained, deep-throated but high-pitched trill, and can only be likened to the sound of one hundred canaries in unison. You are not likely to hear it except where there is enough space and distance to justify its use, though Samba will sometimes trill when she hears children laughing and playing in the street.

A champing of teeth, with saliva falling from the mouth, and much jumping up and down, is nothing much to worry about in the case of the female woolly. If you see it is a male, run for your life.

"Huh, huh, huh" is a good-hearted chuckle, accompanied by rough and tumble, pinches, twists, bites, squeaks, and squeals, with no ill-feeling and no special object in mind.

"Aarrk" is explosive, similar to the swift and short jerk of a rattle. It usually occurs in real horse-play, when the female holds on fast and allows and even invites you to apply a number of hard play-bites and pinches. The motto of this game seems to be: "Please bite me hard so that I can bite you harder!" Her ecstasy is spiced with just a touch of fear, but if you stop she is capable of shoving her foot right into your mouth and demanding more, with a heap of "huh, huh, huhs" to prove she means it. If you are foolish enough to comply, the "aarrks" will develop into "aaaarrrrk", and if you push beyond this you will soon be in a situation that only a male woolly monkey can resolve!

The real measure of Samba's intelligence is shown by her understanding of our language. If she fails to comply with my commands, it is always out of disobedience or because she is too preoccupied to pay attention, but the quickness and accuracy with which she complies when she wants to, show that she fully understands their meaning. The following phrases can be said in a variety of inflexions and tones. They can be made forcefully or casually, or even whispered, and they can be spoken by anyone, though this does not mean she will respond to orders that are unreasonable.

"Take me for a walk." Samba will take you by the hand and conduct you on a tour of the room, or of the whole house if you permit her.

"Don't tail the chair." This is when Samba is out for a game and has gripped the leg or rail of a chair with her tail and is about to pull it over with a crash to the floor. She always releases the tail-hold when given this order. She also obeys "No tail!"

"No biting." Her understanding of this is remarkable, because "no" acquires an absolute meaning for most animals in the course of their training. If I give this order to Liz, who may be biting a book she has picked up, she will drop the book and turn her attention to something else. But Samba distinguishes "no" from "no biting". If I say "no" the book will be placed on the floor. If I say "no biting", she will stop biting but continue to occupy herself with the book, turning it over in her hands and opening the pages.

"Be careful!" Samba has finished her tea and is placing the cup perilously near the edge of the table. Her hand is making tentative movements and she is concentrating hard. I shout "Be careful!" so that she will manœuvre the cup into a safe position. If it just balances, the operation is considered a success.

"Jump on Max!" Our Alsatian dog does not like this, but Samba usually obliges, with plenty of "huh, huhs" and sometimes a "nyonk" or two.

"Go to June." Samba knows everyone in the family by name, and will usually comply, but it is unreasonable to give such requests the force of an order unless it is important. For a dog an order is an honour and a pleasure to obey, but never for a monkey.

Samba is familiar with many more phrases as well as variants of the ones I have recorded. The following are selected from these and I record them without comment: "Want to dance?", "Take off my glasses", "Want to swing?", "Would you like a cup of tea?", "Jump", "Don't open the door", "Don't go in the kitchen", "Come back in here", "Don't go in the lounge", "Stay in the kitchen", "Stay in the lounge", "Don't go in the hall", "Get off", "Time to go, goodbye", "Come on, back to the monkey-room", "Where's your hat?" One can substitute "yes", and "Yes, you can" for all the "don'ts" that appear in the foregoing phrases and Samba will respond accordingly.

Samba can lift a cup and drink from it almost as skilfully as a chimpanzee and replace it on the table slowly and carefully so as to avoid spilling or dropping the cup. She can unscrew the lid from any jar and remove the contents, turn any kind of door-handle and open the door, and will allow her face and hands to be washed when they are sticky with fruit or jam. She can push a go-cart around, sit in it while you give her a ride, or pull it along with her tail. She will look at television with a sustained and concentrated interest for as long as ten minutes, saying "eelk" with great excitement to any animal that appears on the screen, particularly a dog or a monkey. She trusts us when we swing her round and round by the tail, even when she has no hold of her own and her safety depends on us. She never bears a grudge when she receives a hiding for being particularly naughty, such as when she bites Max too hard or tries to twist off his ear.

Max is completely fooled by Samba's sub-human appearance and intelligence. He regards her as a very small and black-faced human being, and would no more think of biting her than he would any other member of the family. He is too noble to be afraid, and when Samba cheats he will sometimes come over to me with an expression that seems to say: "Why *does* she play so rough?" When she gets over-excited and really sets about him, the poor chap is disconcerted and doesn't know what to do. During these tussles, Samba draws upon the woolly monkey's whole repertoire of bad language, with squeaks, "nyonks", and "aarrks", but Max is not allowed to bite, bark or even growl. These are Samba's terms, and if they are unacceptable to Max he must retire defeated. But for the fact that he believes her to be a human dwarf, he would have eaten her long ago.

4

Monkeys in the Amazon

The early literature on animal life in the Amazon regions contains very little information on the behaviour of South American monkeys. The naturalists Wallace and Bates confined their travels mainly to the rivers, making short saunters along the banks in search of butterflies. When we consider the forbidding character and density of the neo-tropical forests, we can sympathise with some of the Victorian naturalists in their reluctance to penetrate more than a few feet into the jungle itself. As for observing the arboreal monkeys who live there, on the jungle roof, the explorer Paul Fountain remarks: "It is extremely difficult to learn about the habitat of monkeys in a vast and dense forest where most of the trees are 150 feet high. If five are seen, possibly there are fifty more out of sight. All the monkeys are shy, and if watched remain still. They are lively and noisy, but avoid showing themselves. I have heard them near my camp for days together without seeing one. Only by concealing oneself and watching patiently can a glimpse of them be obtained."[1]

There are many descriptions of pet monkeys who have attached themselves to the camps, and descriptions of dead monkeys, especially those seen roasting in preparation for the traveller's lunch. Wallace writes: "We caught a glimpse of some monkeys skipping about among the trees, leaping from branch to branch, and passing from one tree to another with the greatest ease. At last one approached too near for its safety. Mr Leavens fired, and it fell, the rest making off with all possible speed. Having often heard how good monkey was, I took it home, and had it cut up and fried for breakfast: there was about as much of it as a fowl, and the meat something resembled rabbit, without any very peculiar or unpleasant flavour. Another new dish was the Cotia or Agouti, a little animal . . ."[2] (A great deal of Wallace's reporting reads like a recipe book.)

Almost every book on the early expeditions to the Amazon contains similar accounts of how to roast monkeys and what they taste like, which are rather disconcerting for one in search of information on the habits of live animals. But not all the Victorian naturalists were like Wallace. Paul Marcoy, who travelled through the Amazon region in 1860, refused to eat monkeys because they reminded him of children. He writes: "A large and black woolly monkey that was tied to the prow of the boat, broke his cord and sprang upon me at a single bound. He threw his long arms around my neck and I had every prospect of being strangled. I struck him on the nose and tried to free myself from his embrace, when I already felt that his arms were beginning to relax. I saw his teeth chatter, and noticed a deep look of poignant sorrow in his eyes, and I was immediately ashamed of having struck that grandfather of humanity. To make what amends I could for my disgraceful act, I kissed him on his black muzzle."[3]

Among the many stories that have been told by hunters, there is one in particular which has often been disputed, but which I am convinced is true. It is said of neo-tropical monkeys that if a young one is abandoned by its mother in the panic of flight, the first female that passes will take it on her back and continue running. I believe this because of Samba's habit of grasping her teddy-bear when she is alarmed and taking it with her on her back to a position of safety. She is never so frightened as to take flight, but often some unexpected and unfamiliar sound will cause her to run a short distance and climb to some high point in the room, usually the mantelpiece or the top of the piano, and when she does she will invariably go out of her way to capture her teddy first.

Very little is known about the feeding habits of woolly monkeys in the wild. Observers have reported prehensile-tailed monkeys as eating large insects, and even young birds and eggs, but no one seems to be sure. Our monkeys are expert at catching moths, and they will eat caterpillars. The speed of their hand movements and the ease with which they capture a moth or a butterfly in flight is a most impressive sight.

It is generally known that most monkeys live to a great age. There was a capuchin on my round at the Melbourne zoo who was forty-two years old at the time I left in 1951. If we ignore the hazards of jungle life, of disease and predatory animals, I think we can say quite safely that woolly monkeys have a life-span of twenty-five to thirty years.*

I have found the published literature on the specialised organs of the woolly monkeys and the method of woolly monkey movement too generalised and vague to have any real value. Before presenting my own observations and conclusions on this subject, a brief comment must be made

---

*As I write today (1980) Lulu at the age of twenty-six is still alive and well. She is now a grandmother, and her son Danny is the leader of our colony in Cornwall.

on the unique character of the vegetation and geography of the Amazon jungles.

*   *   *

The geography of the Amazon region is continually being changed by floods and storms. Even the great rivers shift their channels. Giant trees measuring up to two hundred feet high and ten feet in diameter are uprooted by winds and floods. Whole forests that have reached full maturity disappear owing to the action of rivers, and sections of forests several hundred yards in length have been seen to slide, with a great roaring sound, into the deep and swift-running waters of the Maranon and the Orinoco. Islands ten miles long, covered in dense jungle and teeming with reptiles, birds, and mammals, appear where none existed a year before.

Small birds and lizards are trapped in huge spider nets that sometimes envelop whole trees. A star performer among the poisonous bird-catching spiders is the *mygale*, which can measure seven inches across. Among a rich variety of poisonous ants, flies and hornets, the *isango*, a minute blood-feeding insect, commands a special respect among the Indian tribes.

When the railroad was being built from San Antonio to Esperanza Falls, every inch of the line had to advance through almost impenetrable vegetation. Immense trees cut through at the base remained erect and held in place by the network of vines that tied their tops to everything within a radius of fifty feet. Craig, the chief engineer on this project, wrote: "The density of the forests was such that not even the native Indians could move ten feet without cutting their way through the web of vines; some thorny, others sticky and greasy, and all alive with biting red and black ants of all sizes. These made life one continuous torture. Punctures from palm leaves which lay face down, with myriads of needles standing erect, caused infection and permanent sores. The blow of an axe on a tree trunk would make it rain ants on us by the thousand from the foliage above. The bites of the large black ants caused a terrible swelling."[4]

In this great continental sea of vegetation that has become known as the "Green Hell" it is easy to see why the activity of monkeys and other small mammals is confined almost entirely to the roof of the jungle. The perils of the jungle floor make tree life a condition of survival. For the arboreal monkeys that live there the law of life must be: stay off the ground.

But life in the trees is by no means free of danger, for there are tree-climbing snakes and predatory eagles, and jaguars have been seen pursuing monkeys in the trees at a height of more than thirty feet from the ground. Paul Fountain writes: "Monkeys seldom perceive the approach of their main enemies, the poisonous tree snakes and the jaguar. It is not only colour or similarity to some other object that enables the snake and the jaguar to approach their victims, but silent cunning and craft. The monkey sees not, knows not, until the death-dealing paw pins him down, or the darting head of the snake has struck."[5]

The explorer Thomas Belt writes of an encounter between a monkey and an eagle. The monkey had been heard crying in the forest for more than two hours, and on investigation both predator and prey were found facing each other on the same branch of the tree. Had the monkey turned its back the eagle would have seized it. The monkey, however, kept its face to the eagle, which did not care to attack in this position. How this particular contest would have been resolved was never known, for the eagle was fired at and frightened away.[6]

It is easy to see why woolly monkeys in captivity are readily disturbed by all manner of domestic sounds—the rattling of knives and forks in the kitchen drawer, clock chimes, the scraping of a knife on concrete, even the clattering of the handle of a bucket. In the interior of the swamp forests of the Amazon, the animal and bird sounds are like nothing on earth: howls, clicks, roars, and the shrill screams of eagles and hawks.

With this picture of the Amazon in mind, we are in a position to consider some of the important features of primate evolution in the Amazon, and also the physical characteristics of the monkeys who have survived it.

It is generally accepted that man and many other land mammals crossed the Bering Strait during the Pleistocene period, when there was a land bridge connecting east and west in the northern hemisphere. When the early primates went south to the neotropical forests they evolved into "five-handed" climbers, brachiators, "clingers" and tail-swingers. The atrophied thumb and the hand-tail are therefore reciprocal features in the evolution of neotropical monkeys as a whole, and in order to understand this relation we will consider first the anatomy and the function of the prehensile tail.

The comparative body- and tail-lengths of some of our woolly monkeys are shown in the following table:

Head and body	Tail
Samba:	17 inches	25 inches
Lulu:	17 inches	26 inches
Liz:	18 inches	27½ inches
Pepi:	23 inches	33 inches

Specimens quoted in *Mammals of Amazonia:*[7]

	Head and body	Tail
(A) Average:	450 mm (18 inches)	680 mm (27 inches)
(B) Largest in collection:	705 mm (28 inches)	1,200 mm (47 inches)

It will be seen that the tail-length of the largest specimen (B) is double that of the body, if we allow five inches for the head and neck. With the spider monkey, the largest of the neotropical primates, the tail-length is nearly four feet in length and five inches in diameter at its base! To take Pepi as our example, the tail tapers down rather suddenly from four inches at the base to three inches, maintaining this girth for approximately sixteen inches and then tapering consistently from here to its tip. These facts are important for a proper understanding of the nature of the movement of the woolly monkey in running, walking, climbing, holding, gripping, hanging, clinging on, and leaping.

When woolly monkeys are venturesome and allow themselves to drop four or five feet from a hanging position to the ground, they usually land on all fours in a most ungainly fashion. One could say that they are inexpert at falling or dropping to the ground because they are so expert at not falling! That is why a woolly monkey could never be knocked or pushed off the branch of a tree. It would have to be scraped off. Hunters in the Amazon have in fact shot them and found them still hanging by the tail when they are dead. Even in death, it seems, the instinct to stay in the trees lives on. How difficult it must be for a predatory eagle to capture even a young and small woolly monkey! A woolly monkey attacked will only take to flight when the likelihood of escape is almost certain. If there is the slightest doubt, it will cling on and face the enemy. This is characteristic of all our monkeys in their play activity. The anatomy of the woolly monkey is ideally suited for clinging and wrapping itself round an object. When any of our adult monkeys do this during play they cannot be dislodged. It might be possible to detach them if one were to apply all one's strength, but only at the risk of injuring the animal.

Woolly monkeys can and do brachiate (a hand-over-hand movement), but the tail is in constant use no matter what form of movement they adopt. Brachiating is not always the best form of movement in close canopy jungle. In no other jungle is there such a prolofic growth of vine as in the Amazon, or so much vegetation that is decomposing and ready to break with the weight of an arboreal animal, even near the roof of the jungle. The instinct for tree security is so strong in the Amazon monkeys that all five "hands" will be fully employed even in positions of complete safety, only a few feet above the ground.

When woolly monkeys brachiate, it is always with the added security (not necessarily support) of the tail. Even when it is not being used with specific advantage, it will add its security to whatever is going on. If you carry a woolly monkey for a walk round the garden, the tail will always be wrapped around you. The woolly monkey will hang by one arm, like the gibbon, but you will never find the tail hanging or waving about in space as though it were a superfluous organ, an impression one often has when watching the long-tailed monkeys of the Old World.

Woolly monkeys never hang by the tail alone unless the supporting branch has proved strong enough to take the weight. When they are satisfied, from long experience, that a particular branch or beam is secure, they will perform a number of amusing tricks in the tail-hanging position. They will also swing on a long and isolated rope, but not until the rope has been tested. Suspend a new rope from a great height, and see how tentatively they approach and investigate before climbing or swinging on it. Our monkeys will allow any member of our family to pick them up and swing them round by the tail, but I would never allow a friend to do this, so fearful am I of losing the monkey's confidence should there be an accident. A woolly monkey, strangely enough, has little objection to having its tail held, but immediately a stranger attempts to lift him by it he will leap up and grab the offending arm.

There is another species-preserving function of the tail of particular interest, one that I have seen illustrated in a natural as well as a domestic situation. When Samba is in doubt she investigates first with her tail. In the garden I have often observed her occupied with some strange new object, a staghorn beetle, say, or a toad. Standing on two legs with her back to the toad, while peering round for a complete view, she advances cautiously and tests its reaction by a whisper of a touch with the tip of her tail. When there is no repsonse from the innocent toad, the tail becomes more venturesome, turning it over and prodding it to see whether it bites. If it moves, Samba will jump up and down with excitement and uncertainty. As her confidence increases the toad is man-handled, until it has to be rescued from certain death.

The logic of Samba's approach to the toad is clear: she does not want to lose the tip of her tail, but it is, nevertheless, the least vulnerable part of her body. Her method of investigation also keeps her body twenty-five inches away from possible danger, and leaves her hands, feet, and mouth free for action. The same ritual is performed in the house when she is confronted with a strange object. On one occasion she disliked and therefore "yooked" at a toy giraffe that was on the floor in an upright position. It was a repeat performance of the encounter with the toad, only this time she also acted the part of the toreador by grabbing a cloth-cover that happened to be near by and throwing it over the giraffe. This would seem to be an approach to the stage of a tool-using primate. The technique of combining the use of the tail with a large palm-leaf could be very effective for dealing with small and unwelcome visitors in the Amazon trees.

I have indicated that the main function of the hand-tail is for hanging, gripping, and clinging, as well as for general movement; it is now important to relate this to the woolly monkey's atrophied thumb, for the atrophied thumb and the prehensile tail must be recognised as reciprocal features in the evolution of hand-tailed neotropical monkeys as a whole.

In swinging slowly through dense foliage, the thumb obviously becomes a hindrance. It is less of a hindrance for fast and complete brachiators, like the gibbon, but even the gibbon, compared with the anthropoid ape, has a relatively small thumb that is set well back in the hand, away from the other fingers. In their own habitat woolly monkeys spend a great deal of time simply hanging. They are "hooked" in the trees as it were, a situation in which the developed opposable thumb is less advantageous than an atrophied thumb which functions like the little or fourth finger. Now a tree-dweller, with or without a tail, does not advance through the tree backwards, nor does he advance forwards with his feet! Neither does a monkey brachiate upside down with his feet; at least, not outside a circus. In full or partial brachiating, the feet bring up the rear, so that the opposable thumbs of the feet are ready for a secure grip in any upright or semi-upright position that may be necessary. The opposable foot-thumb therefore (or great toe if you insist), far from being an impediment to moving or grasping, becomes a decided advantage for any movement that is non-brachiating.

And so we find, logically, that the foot-thumb of the woolly monkey is extremely well developed, even *more opposable* than the thumb of the human hand, and well separated from the other fingers. If the foot-hand of a woolly monkey were enlarged to the size of the human hand, it would stretch nearly two octaves on the keyboard of a piano. I have been careful to say, if the *foot-hand* were enlarged, for woolly monkeys have enormous hands and fingers in relation to their size, if the *monkey* were enlarged to the same size as a human being, its foot-hand would be able to stretch at least two and a half octaves, or approximately sixteen inches!

The hand-thumbs of the woolly monkey are usually described as being "well-developed". This is misleading. It is true that of all the neotropical hand-tailed monkeys the species with the least atrophied hand-thumbs are the capuchin and the woolly monkey, but compared with the well-developed and opposable hand-thumb of the Old World monkeys, such as the Indian rhesus, that of the woolly monkey is poorly developed and, moreover, is not even opposable. The atrophied thumb has the appearance of a fourth or little finger, but has less power and flexibility. When a woolly monkey picks up objects with the hand (remembering that it frequently uses the tail or the feet for picking up) the thumb moves in the same direction as the other fingers. When Samba unscrews a lid from a jar or applies herself to any action which requires twisting or turning she will often use her feet, for the foot-thumb is opposable and well-developed, as it is with the orang-utan, who is also four-handed. Expressing it in common language, the hand-thumb in the evolution of South American primates is "on its way out".

We have seen that an accompanying characteristic of the hand-tail is a less developed opposable thumb (as in the case of the capuchin), or an atrophied thumb (as with the woolly monkey), or no thumb at all (as with most spider monkeys). And it is precisely where the thumb is most atrophied, where it is reduced to a mere stub or where it is entirely absent, that the prehensile tail reaches its fullest and most magnificent development—in the spider monkey.

It may be argued that there are arboreal mammals in the Amazon who have managed to survive quite well with no tails, or with short half-tails like the uacaris. The point is that in the evolution of the hand-tailed primates there is a development towards the large and strong prehensile hand-tail, as found in all species in the group Ceboids, the most developed of the Amazon monkeys, as distinct from half-monkeys or Pithecoids. It is also true that there are many other mammals of lower evolutionary descent, such as the harvest mouse and the opossum, who have prehensile or part-prehensile tails, but very few of them are able to support themselves by hanging from the tail alone, and none of them can use their tails in the strict sense of "hand-tail", or with anything approaching the technique of grasping, holding and throwing that we find in the group Ceboids. (Samba often picks up a plastic bowl and throws it at me with her tail, while she is swinging on a rope.) Their nearest rival is the kinkajou.

Although some writers seem to be hypnotised by special-ised features, they forget that the use-value and function of a special organ such as the prehensile tail—wonderful though it is in its own right—has only a restricted and particular value for a given species in a given environment. Yet this has not prevented them from speculating on the value of certain specialisations, such as the hand-tail, had they become universally adopted. I sometimes think that the gorilla would have become a greater object of interest for them had it been blessed with inverted nipples.

Nature is not quite so foolish as to have evolved an organ as intricate and sensitive and of such tremendous strength and size relative to the rest of the animal as the hand-tail, unless the survival of the animal in a given environment depended upon it. When we take the broader view—of the evolution of primates from monkeys through the apes to man—the tail itself can only be seen as a hindrance. It would be no less of an embarrassment for the chimpanzee than it would be for a primate concert guitarist, and God knows what kind of tail would be necessary to support a gorilla. Conversely, only the Old World provides the necessary conditions for the full evolution of primates—from half-monkeys, monkeys, baboons, gibbons, anthropoid apes, to mankind. The ground-living patas monkeys of Africa, to take an extreme example, are found in the open savannahs and grasslands. They run and walk almost with the gait of dogs, moving on the flats of their hands and feet.

Many of the Old World monkeys not only are good climbers and well adapted to arboreal life, but can move on the ground much faster than prehensile-tailed monkeys. Old World monkeys are also faster and more agile at running along branches and up and down the trunks of trees. But if the ground-living patas monkeys were to be dropped into the Amazon jungle they would quickly revert to an almost exclusive tree-life. And if they were able to survive for a million years or so they would probably develop prehensile tails.

The evolution of the hand-tail primates can only be

understood in relation to the special problems of survival presented by the vegetation and structure of the Amazon, where the *evolution of the jungle itself follows a different pattern*. Even in the remote parts of Malaya, Sumatra and Borneo, where the jungle terrain and flora may be compared with that of the Amazon, the problems of survival are different in character.

If I were going to explore one of the habitats of the woolly monkey in the Amazon, and Samba could talk, I am sure she would give me the advice a Peruvian bushman gave to the explorer Leonard Clark: "White men look only ahead of them, but you must look even closer ahead, much closer!— and above, as well, and below, and on both sides, and especially behind. Never forget to look behind!"[8] And I am sure Samba would add: "Don't forget to take your prehensile tail—above all, you will need that!"

5

Character and Dominance

In my view, a study in woolly monkey dominance must be related not only to the character of the individual monkeys, but also to different character types. In our private language at home we have always distinguished two main character types by the terms "long-bodied silver-greys" (group 1), and "stocky dark-greys" (group II). Both groups belong to the species *lagotricha lagotricha* of the genus *Lagothrix*, and represent a variation of type within the species. In our community of six monkeys, Pepi and Liz belong to group I, and Lulu, Jess, Jimmy and Samba to group II.

Group I monkeys—"long-bodied silver-greys"—are more aggressive and less relaxed. They are continually on the alert, with a fixed expression about the eyes of "rhesus-like" or typically "monkey" alertness. Their play-bites are uncontrolled and painful to bear, and by nature they are less reliable. They are less intelligent about what they eat, and will often make themselves sick from over-eating. They tend to be somewhat larger animals than the monkeys in group II.

The monkeys in group II—"stocky dark greys"—are more amiable in disposition, more responsive to human influence, more reliable and predictable, and less aggressive. This does not mean that they are less dynamic or over-civilised; they are in fact the dominant members of the monkey community. They are more individualistic, more expressive, and as fast in movement as the others when they want to be. They move from *largo* to *presto*, whereas group I monkeys are almost always *allegro*. They are not so "greedy" in their food habits. The relaxed and subtle shades of expression that flicker about the eyes of the "stocky dark-greys" remind one of the gorilla. In many ways they are like miniature gorillas.

Although these distinctions in character type are un-likely to receive zoological classification, I nevertheless believe that they are important, and I therefore ask the reader to keep them in mind.

*   *   *

In woolly monkey society there is female as well as male dominance. There is also a social order of rank that embraces females and males alike. In our group of six monkeys there are two adult males, Pepi and Jess; two adult females, Lulu and Liz; a three-year-old female, Samba; and eighteen-months-old Jimmy. The order of rank was as follows: Jess, Pepi, Lulu, Samba, Liz, Jimmy.

Jess was a mature adult when he first joined our monkey community, and during the one year he was with us his position as ruler was assumed rather than asserted. He acted like a benevolent emperor from the moment he arrived, and his position was never disputed. It was difficult at first to explain the respect he commanded from the others, for there were many reasons why this was unexpected.

In his previous home Jess had lived as a pet, enjoying the freedom of a large garden with trees. He was five years old when he came to us, a strong, well-developed and healthy animal, who settled down in his new home quickly and happily and was friendly and affectionate with everyone. (I had hopes of him as a mate for Samba, but he died from enteritis a year later.) He displayed no sexual interest in Lulu, who had already mated with Pepi, nor in the other females, Liz and Samba. His experience of monkey communal life was nil. He did not fret for his previous master, and he was on good terms with everyone, monkeys and humans alike. Even Pepi, a stronger, larger, sexually mature animal, and highly possessive about Lulu, accepted Jess as ruler.

Jess may have been sexually immature, for most male woollies are already performing the thrusting actions of copulation at the age of three, and Jess was well into his sixth year. His sexual development may have been retarded from want of stimulation and example since he had lived his five years almost exclusively in human society. Whatever the explanation, he was not a negative or submissive animal. He had been disciplined, but his individuality and his monkey-ego were as strongly marked as those of the other monkeys.

I have a habit of working late through the night, and many were the nights he spent with me, walking round my typewriter, waiting for the bell to ring and responding with "eelk", helping himself to my tea, and eventually falling asleep by my side when he was tired. Sometimes he would sit up, yawn, scratch his stern, and return with one of his subjects. Any one of them would do: Liz, Jimmy, or even Pepi, for Lulu was not always disposed to spend the night with her master.

Samba was interested in Jess, and would often present herself to him with unmistakable intent, but Jess always treated her like a wise old man who understands but is unwilling or unable to oblige. Another extraordinary thing was that Jess played with Lulu, and Pepi did not object. Jess was the least aggressive of all the monkeys. Although a powerful animal, he was not as strong as Pepi, and he was five pounds lighter in weight. He took no special interest in Liz or Samba, both of whom were available and never shunned him. He took no special interest in anyone. Why was he almost *encouraged* by the others to be the leader?

Pepi had protruding canines that overlapped and interlocked. Jess's canines were small by comparison, and no larger than Samba's are today. I had rough games with Pepi and Jess, and I assure you that Pepi was much the stronger animal. Pepi's play-bites were unbearable for a human, unbearable for Liz and sometimes even for Lulu, but I never once heard Jess complain. Whenever I gave Jess a hard blow, for he was often deliberately naughty and seemed to enjoy the punishment, he would chuckle with glee, turn his rear end round and poke at me with his tail, just as a conscientious mother will test the baby's bathwater with her elbow to see whether it is safe.

Jess was very good at in-fighting. In a play-fight with Pepi he would always drop his head in the first attack and allow Pepi to concentrate on the back of his neck. (The thick section of muscle and bone that joins a woolly monkey's head to his back can hardly be called a neck! It is the most vulnerable part of his body.) While Pepi was occupied, Jess, who had grown up with a large bull terrier, would work hard at Pepi's feet, hands and tail. It was always Pepi who retired, running away, as Jimmy would run away from him, defeated and humiliated.

Pepi's fight sounds of "aarrk" were double the volume of Jess's, but his displays of defiance and intimidation, calculated to alarm Jess, had no effect at all. When Pepi, drooling with saliva and flashing his canines, made an awe-inspiring leap, Jess would sit there, his arms stretched out and ready, waiting for the clash that never came, for Pepi would always break his leap with a high and clumsy back-jump, and make off, careering round the garden with Jess hot on his heels.

Jess was indeed a peculiar fellow; he was also a gentleman. In his games with Liz, who is the first to complain when hurt, he never once caused her to squeal with pain. Samba and Liz were rarely seen playing with Pepi or Lulu, but they loved a game with Jess. And Jess was the only fully grown *male* monkey we ever knew who was never spiteful with children. During the first three years of his life he had been a photographer's model and gimmick, and had learned to tolerate almost anything. He should never have been a monkey.

Can you imagine five monkeys, including Pepi, all waiting patiently round the food-tray, while Jess picked out the delicacies in his own leisurely fashion? I only once saw Liz reach out and very quietly help herself while Jess was still at the tray. Jess was quite unconcerned. I know that he would not have objected had the others joined him at the meal, for by nature he was not greedy or possessive.

When Jess died (from a virus infection in 1965), Pepi became the leader in a community of five monkeys. He was a despot. Intruders or interceptors at the food-tray were not to be thought of as he settled down to a great feast, while four peckish monkeys waited and "yooked" their impatience. Only Lulu was allowed to join him, and even she would sometimes wait until his lordship had finished.

Pepi was twice Lulu's weight, and, unlike Jess, he was no gentleman. But Lulu was the strongest and fastest in action of all the females, and in her play with Pepi she fought back like a tiger, Lulu was a girl who enjoyed being pursued and captured. Her problem with Pepi was to maintain a tempo of flight consistent with the thrill of escape, yet not so fast as to make it impossible for Pepi to catch her. It was like watching a weight-lifter chasing Charlie Chaplin. Of course, Pepi could never catch Lulu by his own initiative. Lulu would cover thirty feet of territory, crossing horizontal beams, sliding down poles and racing up again, while Pepi lumbered along behind, making the most impressive, giant-like leaps, but always arriving too late. The only way Lulu could make these pursuit-games interesting was to be adventurous and foolhardy. During one of her most exciting and daring leaps, I saw Pepi swing past her in the opposite direction; they both finished at opposite ends of the enclosure, Lulu seated and composed, and Pepi hanging by one arm and his tail, looking angry and bewildered.

Lulu was the finest gymnast of all our monkeys, males included. She could swing thirty feet on the end of a rope from a height of fourteen feet and back again, landing with the grace of a bird. No doubt she could have trebled the scope of this performance in the Amazon jungles. All our monkeys are good climbers, acrobats and rope swingers. Over the years the space and the equipment have always been at their disposal. But there was only one Lulu. In her, skill reached such a high level that one forgot technique and dwelt only on her art.

The time came when Samba began a struggle for dominance with Lulu! Samba at this time was a little more than three years old, and Lulu was nearly five. Lulu was therefore older, stronger, bigger, more experienced, and the undisputed ruler in the female hierarchy. Among the many encounters I witnessed during the first few months, two fights were serious. Needless to say, Samba lost both of them. In the first fight the end of her prehensile tail received a two-inch gash that took a week to mend. For several days she did a minimum of climbing, and spent most of her time sitting on the window-sill of the indoor monkey-room, looking ruefully through the glass at the others while they played outside in the sun. She spent the evenings with us in the house, keeping her tail well out of harm's way, especially when she sat down. Monkey wounds heal very quickly, and after the fourth day she began to use her tail again. During the period of recovery, life in the monkey community continued without incident. Lulu, who was never the first to begin any quarrel, behaved as though nothing had happened, and on the first day that Samba ventured into the garden enclosure greeted her with "eelk".

The second serious fight occurred about a month later, and I was there when it began. Samba was sitting by the hatch door of the runway that connects with the monkey-room and the garden enclosure. Lulu, returning from the enclosure, came down this runway with the object of re-entering the house through the hatch, but Samba made no attempt either to move out of the way or to pass through the hatch so that Lulu could enter. Lulu could hardly be expected to place herself at Samba's disposal, and so in her customary elegant fashion she edged past Samba, pushed up the hatch , and was about to enter, when Samba reached out and grabbed the end of her tail. Now this was sheer provocation. Lulu and some twenty-five inches of tail had already passed Samba, and it was the end of the tail that Samba grabbed before it disappeared through the hatch! It was not a play-gesture, there was no "huh, huh" and Samba's face was inscrutable. Lulu turned and made a threatening gesture, and in seconds they were locked in battle. In all such combats, whether mild or serious, there was never any question of flight or pursuit on the part of either monkey. Samba's method of retiring was to disengage herself when she had had enough, and scream when Lulu administered the deciding bite. She would then sit back, a few feet away, examining her wounds and "yooking", while Lulu walked off with dignity, and without visible signs of injury.

The result of the second fight was for Samba more serious. A deep tear-bite across the sole of her foot put her out of action for another week, and this time she had to walk with one foot off the ground. When the wound had healed, for many days she walked from habit on the side of the injured foot. And even months later she would lift up her foot for inspection and for a continuation of the sympathy and consolation I gave her at the time of the injury.

I should mention here that in the intervals between their serious encounters Lulu and Samba often engaged each other in friendly "huh, huh" play-fights. These rough games had an entirely different character, and though they were all resolved in Lulu's favour, they ended with no hard feelings on either side.

Eventually the day came when one of the minor bouts to my great astonishment seemed to end in a draw, for both monkeys sat a few feet apart "yooking" at each other. I had never known Lulu to "yook" before, except from fear of thunder and lightning, or the new moon, or because a large bird unexpectedly flew across the outside enclosure with a great flapping of its wings. During the weeks that followed, Samba gradually climbed to a position of equal status with Lulu in every aspect of life both in and outside the monkey community.

The perplexing thing about Samba is her persistence. Why has Liz, the largest and strongest of all the females, remained contented with her rank, the lowest in the order?

Why has she never made a bid for dominance? Even little Jimmy, half the weight of Liz and with no more than one third of her strength, is nearly her equal. And what induced Samba to persist in her rise to dominance?

Liz is always the first to squeal in a rough game, and like most squealers she is the first to break the rules. Her play-bites are exceedingly painful to bear. Watching her at play with Samba, I am forced at times to regard her as a damn fool. She asks for trouble, and when trouble comes her screams are enough to raise the dead. For all her subservience to Samba's rule, they remain great friends. Together with Jimmy they often cuddle up with the mutual sobbing ritual, sleeping together at night and spending a great part of their day in fun and games. But when meal-time arrives stark realism prevails. With hands, feet and mouth and sometimes tail, Liz seizes the delicacies and makes off with them to safety, though not always successfully. A leap from Samba and a scream from Liz—and she is bereft of her ill-gotten gains.

I can only think of one explanation of the rise to dominance of Jess and Samba. It would appear that under certain conditions a smaller monkey is able to convey its strength of character to the others. I do not mean in some mystical or anthropomorphic sense, but as evidenced by the power of Jess and Samba to *withstand more punishment* and by their superior intelligence. For similar reasons, Samba has been able to dominate our Alsatian, Max, whereas Liz is terrified and will have nothing to do with him. A similar comparison can be made between Jimmy and the newest arrival in our monkey community—Django, a young male who is larger and stronger than Jimmy. Django's play-bites are far stronger than Jimmy's, but Jimmy never complains. Django, however, squeals frequently, and when I have occasion to bite him (which is the most effective way of disciplining an unruly monkey) he screams with rage. When I bite Jimmy he usually laughs and bites back, and I am so demoralised I haven't the heart to find out just how hard I would have to bite in order to cause him real distress. Django is unquestionably the stronger monkey, but Jimmy is dominant, and in my opinion he will always rule Django because he is able to take more punishment.

Good-natured and courageous dogs just cannot manage monkeys. Such dogs play the game, whereas a cowardly dog, if it is big and strong enough, will kill a monkey. Samba and Jess were both quick to discover the weak spot in the character of the domesticated dog—its inbred humility. If we add to the natural shrewdness and instinct for dominance that drives the monkey-ego the superior character and intelligence of the group II monkeys, it is easy to see how Samba and Jess were able to assert their authority over animals who were much larger and stronger than themselves. Not all monkeys have enough intelligence to do this or to approach new problems (or new dogs) with caution. Samba will have nothing to do with a strange dog. She will bide her time until he proves himself to be either a coward or a sportsman. When a strange dog is brought into our house you will always find Samba on the highest beam in the far corner of the monkey-room, well out of sight, and there she will stay until he has gone. Liz, however, will grab at the stranger through the chain-link of the outside enclosure, and scream with fright should he growl and leap up at her. But if the visitor were to stay with us for a few weeks and prove to be a real dog, a sportsman, you would be likely to find Samba pulling him along by one ear, while Liz was still screaming and poking at him as she swung from a rope beyond his reach.

6

Instincts and Proverbs

"From each according to his ability, to each according to his needs" has been one of the first principles of socialist philosophy. We see it carried into effect, with slight variation, in the monkey community—when Samba eats the jam, butter and middle section of the bread, and discards the crust for Max.

It has often been said that monkeys are wasteful with their food. This is nonsense. In the first place monkeys are wasting *your* food, not theirs. Secondly, they eat and function on the basis of their own proverbs, which are altogether different to human ones. Here are four popular monkey proverbs: 1. "Waste not stomach space on carrots when you can eat grapes." 2. "Only a fool eats what is placed under his nose." 3. "Leave the worst until last; you may not want it." 4. "Never save for tomorrow what you can eat today."

Unlike humans, monkeys keep to their proverbs. Let us consider, for example, proverb 1. If a cat is chewing a piece of stewing beef and you offer it fillet of steak, it will not let the beef fall from its mouth and pounce on the fillet. There may be exceptional cats, but the cat in general will always finish what it is eating first. Monkeys, however, are never so carried away by what they are doing that they are unable to make or accept a change for the better. Only a half-wit monkey would be so foolish as to finish an apple when confronted with grapes. Not only will the apple be dropped unceremoniously, but whatever is in the monkey's mouth will also be expelled.

"Only a fool eats what is placed under his nose" is demonstrated in the following situation. When a food-bowl is tipped up on the dining shelf in the monkey-room, everything is turned over, examined, and selected according to preference. And if you pretend to pop something into your mouth that they are not supposed to see, and stand there chewing in front of them, nothing in the food-bin will be touched. They will climb up, stare at your mouth for some visible signs of what they are missing, and—if you allow them to—even open your mouth, peer inside, and help themselves to whatever takes their fancy. For a monkey what the eye can't see, the belly most certainly can miss.

"Leave the worst till last; you may not want it" is exemplified by the contest as to who can eat the most grapes. Monkeys are extremely logical, and the contest therefore *begins* with who can eat the most *juice*. Each grape will be pressed in the mouth and ejected at lightning speed with a certain amount of the flesh attached to the discarded skin. The skins will be stripped down later, but not if there is something else on the menu that is more attractive.

There is no specific situation to illustrate proverb 4. It works all the time. Monkeys do not save for a rainy day; they have even more faith in Nature the provider than most humans have in their insurance companies. Perhaps the nearest Samba ever comes to having a thought for the future is when she takes an apple to her sleeping box and puts it down beside her without attempting to eat it. This may be because she is vaguely aware that it will provide a useful snack in the middle of the night, or because she intends to eat it immediately she arrives at the sleeping box, but changes her mind when she gets there. I think the latter explanation a more likely one.

*　　*　　*

Discrimination, curiosity and the desire for change are allied characteristics in monkey behaviour. Monkeys thrive on variety, and it is cruel to deprive them of it in captivity. If, for example, raisins are first preference in the daily food supply and shortcakes are last, you have only to provide a surfeit of raisins and make a drastic cut in shortcakes, and the order of preference will be reversed. Provide a surfeit of any favourite item and in time the monkey will tire of it. Reduce to a minimum or remove entirely any item which has been accepted indifferently or fitfully, such as carrot tops, and these will be greeted with enthusiasm when returned a few days later.

I am not suggesting that monkeys have no stable preference for particular foods. The most effective way of discovering what it is would be to exploit the variety factor— provide a surfeit of everything and observe the order of preference over a given period of time. Ignoring the impracticableness of this method, I am still not convinced that it would work, because monkeys are able to change their own minds and habits regardless of the laws of supply and demand, whether in the wild or in captivity.

The need for variety seems to be an integral part of higher primate psychology. It is seen in the human passion for ever-new fads and fashions, and even in the jargon of behavioural science, in which the play-activity of animals is now referred to as "occupational therapy". (The monkey may provide occupational therapy for you, but whatever you provide for the monkey is at best a poor substitute for his own way of life in the wild. I object to the term because it is patronising as well as pedantic.) Variety is particularly important in play-activity. It is a great day for our monkeys when I throw a large cardboard fruit-box into the monkey-room. They climb inside, fight over it, carry it to the top beams and tear it to ribbons. One box will keep them active and amused for an hour or more, but if a box is given to them every day they eventually lose interest. If there is nothing better to do, they will toy with it perfunctorily. But if plastic bowls, a football, rubber boots, old clothes, boxes and bundles of newspapers are each introduced at different times and for varying periods, the interest never wanes.

Our monkeys wake up in the morning very hungry, but not so hungry as to be unconcerned about where they are going to eat. In the winter months they will often bring their food from the monkey-room into the kitchen, where they can enjoy the social atmosphere of the family breakfast. On a summer's morning Jimmy can be seen carrying an apple in his tail, a banana in each hand and a lettuce in his mouth, while he hobbles with great difficulty across the monkey-room, climbs a rope, opens the hatch and passes through the runway that connects the house with the outside enclosure to settle down on the grass for his breakfast in the sun.

*   *   *

Woolly monkeys are no less cautious than they are curious. When we repainted our kitchen, the monkeys were confined to their own quarters. Normally they are free to leave by a hatch that leads into the kitchen, where they often join us, especially at meal-times. (The interior walls of our house are full of holes, with swinging hatches for the convenience of the monkeys. So far the house has remained standing.) The painting was finished in one day, and when we opened both hatches to let them through Samba peered through the kitchen hatch, took one look at the new paint-work (the colours were as before), "yooked", slammed down the hatch and returned to the monkey-room. When I arrived in the monkey-room with the object of reassuring her, she had already left by the outside hatch (another hole in the wall) and was seated on a top beam in the garden enclosure, some forty feet away. She re-entered the house later, but it was not until the afternoon of the next day that she found enough courage to pay us a visit in the kitchen.

I have noticed that human beings who are intensely inquisitive are usually full of mistrust. Like monkeys, but without their justification, they probe and peer into every-thing, registering great astonishment and incredulity over the most elementary facts of life and even looking like monkeys. I think the philosopher Hegel must have had these people in mind when he said: "It is not the business of human beings to regard new objects with the astonishment of animals." We do not, however, expect monkeys to take things on trust; if they were not inquisitive as well as hard to convince they would have become extinct long ago. In the monkey com-munity caution and curiosity go hand in hand, and both characteristics must be regarded as marks of high intelligence.

The innate sense of curiosity in monkeys is more closely related to their instinctive behaviour than is generally realised. We are tempted to regard a number of their play-actions as a sign of "pure" intelligence, when the truth is that the intelligence they display in the investigation of new objects is just as much an expression of a species-preserving function as the prey-catching movements seen in the play-activity of cats. In another chapter I have argued that instincts should not be treated as though they were a blockage to new knowledge and experience, but as ground on which new knowledge and techniques can be developed. For the present, however, I am concerned with the special character of the relation between woolly monkey curiosity and instinctive behaviour.

There are two main instinctive drives behind the monkey's investigation of a new object: is it dangerous? is it edible? Cupboards, boxes, and jars must all be opened; inside may be something very tasty, or very dangerous, or both tasty and dangerous; one never knows. The monkey's chief concern is to find out what is underneath, behind, and—most important of all—*inside*.

In the natural habitat of the woolly monkey, nuts and fruits must be opened, for the best parts are inside. The yolk lies inside the egg. Grubs are found inside holes or under the bark of trees, and they must be poked out. Large insects wear their bones on the outside; they must be pulled apart to be eaten. Dangerous spiders and insects must be rendered harmless from the inside. The monkey and the surgeon have much in common; both are dedicated to the inside, and neither is likely to be led astray by outside appearances.

Whereas the mongoose will put everything into the box, the monkey will take everything out of it. If the mongoose must take everything home, the monkey must take every-thing apart. Our mongoose Nicky and Jess were great friends. They would even lie together on our window-sill in the lounge, basking in the sun. When Jess was absent Nicky had no interest in his whereabouts, but Jess was never happy unless he knew where Nicky was and what she was about.

Jess's curiosity got him into a great deal of trouble. He had a habit of lifting up the cover of our divan and peering underneath to see what Nicky was doing. This was a dangerous thing to do, because a mongoose does not like to be disturbed, especially when it is in the dark and out of sight. But Nicky was good-tempered (by mongoose standards) and had learned to accept Jess's intrusions as part of a game, a game in which she was always the victor. She would run from the back of the divan round to the front, steal up behind Jess and nip him in the rear. Until I actually saw this done, I would never have believed a mongoose capable of such premeditation. It was an incredibly funny event to witness, because woolly monkeys, when they are bending down and peering, have a habit of keeping their tails high in the air. To catch a monkey unawares is one thing, but to surprise him with a nip in his most vulnerable spot is a stroke of genius. Only a mongoose could do it. The speed and the silence with which Nicky left her box under the divan and reappeared on the other side at Jess's stern constituted a miracle that Jess never really accepted. It was outside the realm of monkey possibility.

'. . . here it seems that the tail plays with the monkey' (p. 14)

Outdoor enclosure

Liz on the wheel

Samba

Django climbs his first rope in England

. . . and Liz brings him down

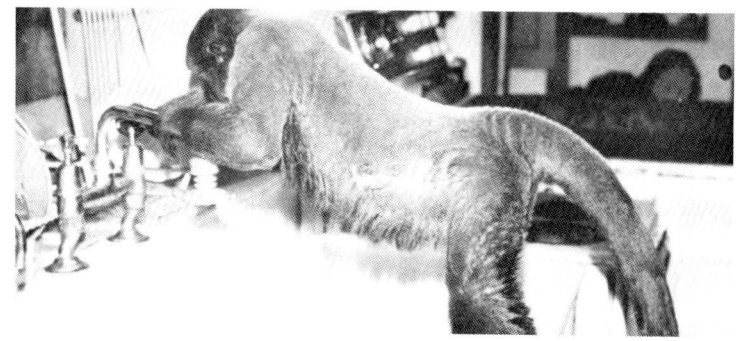

Samba turns on the tap to get her own water

(Bottom left) *Guilt gesture: 'Have I done the wrong thing?'*

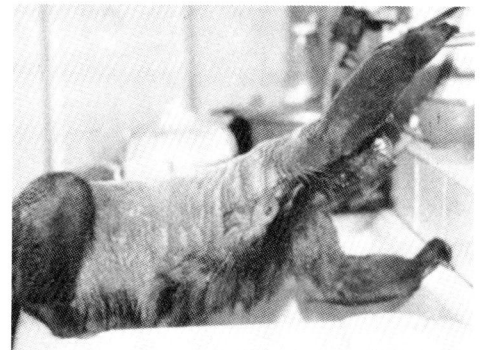

(above) *Samba laughs*

(below) *Jimmy waits to be groomed*

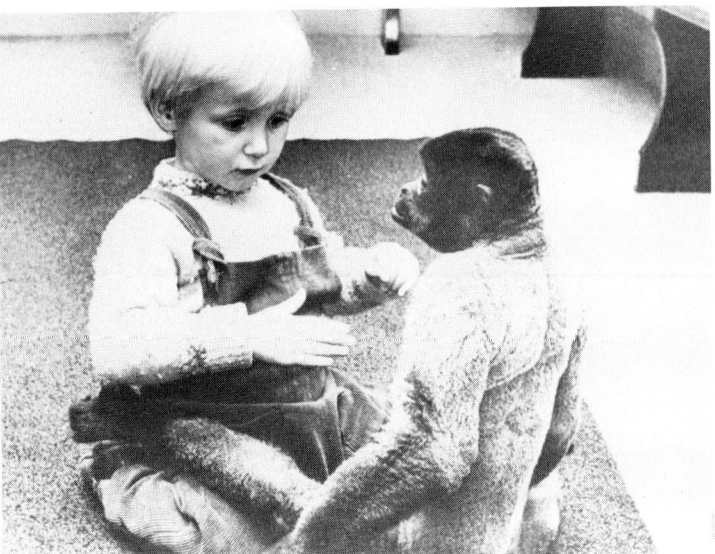

Max and Daniel

'. . . a dull evening at home'

(above) *Jimmy's favourite television programme*

(below) *'Now let me see'*

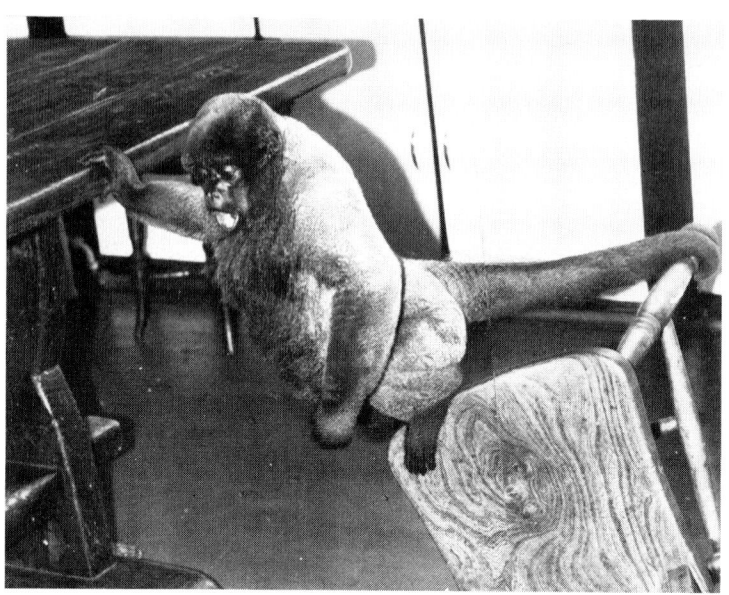

The chair game

*Teddy is rescued from strange toy dog, and cloth
is picked up ready for investigation*

Cloth is dropped on dog to test reaction

Jimmy likes a swing round

June and Liz

At the Monkey Sanctuary in Cornwall

*Indoor
social
activity*

Leonard Williams and friends

Dan with three monkeys

'Why do you attack donkeys?' (p. 114)

Jessy with her first baby at the Monkey Sanctuary

7

Samba Laughs

Intelligence tests and the tricks of performing animals are on the whole very dull affairs, and little is to be learned from them. I believe that we can really enjoy the company of our animal friends only if we allow and encourage them to develop their own individuality. Monkeys in particular must be given the opportunity to invent and solve their own problems and to develop those skills which come naturally to them, for only in this way can their individuality be preserved.

Any kind of training which dominates the monkey-ego will only arrest the development and reduce the health and vitality of the animal. This is not to say there must be no discipline, for without discipline there can be no freedom; the undisciplined monkey is either confined to a cage or tied to the end of a chain. Intelligent animals, like children, soon learn to respect discipline, and it is my experience that the charcter of the monkey is actually strengthened by it.

In the early days, it was necessary to discipline Samba for breaking the children's toys, but afterwards I would always call her back to the scene of the crime, reassure her with an exaggerated display of affection, and ask the child to be kind and return the toy to Samba. In time she learned that there are ways of getting what one wants other than by taking without permission, she learned also to ask, and she learned the virtue of waiting. And whenever one of the children tried to take something away from Samba (not that they ever succeeded, for it is exceedingly difficult to take anything away from a determined monkey) she was delighted when we intervened on her side and saw to it that justice was done. The kind of treatment not only had a civilising effect on Samba, it inspired her to new heights of endeavour and monkey cunning. It did not crush her desire to be naughty, it simply encouraged her to be naughty in a more civilised and sophisticated way. She learned that if you must pinch something, then at least do it nicely and you may get away with it. Like the professional safe-breaker, Samba learned the importance of extreme care and absolute silence.

If a monkey picks up your whisky behind your back, takes a quiet sip and then puts the glass down again very carefully, how can you possibly object? But if a hooligan tears across the room, grabs a handful of porridge from your plate and spreads it from the kitchen table to the top rail of the curtains in the lounge, it is difficult to remain tolerant and placid. If our monkeys want to throw things about and tear them up, they have their own room and their own garden to do it in. They can also urinate wherever they like in their own living quarters, but not in the soundhole of my guitar.

Animals often exhibit moods and responses similar to those characteristic of human behaviour. The similarity may be difficult to see, for in animals the expressions of emotions which correspond with those of human beings are often exaggerated or grotesque by comparison, or so slight and fleeting in nature as to pass unnoticed. When our woolly monkeys jump up and down with joy and excitement, to the inexperienced observer it looks like a war-dance; and when they are tickled under the armpit, their laughter is no more than a whispered "huh, huh, huh" with teeth bared and much head-shaking, so that the observer begins to wonder whether they are angry or in pain.

In his classic work *Expression of Emotions in Man and Animals*, Darwin was primarily concerned with the physical anatomy of the emotions, and although his study is rich in parallels between animal and human emotional behaviour, the emotions considered are necessarily primary and fundamental, such as those of joy, amusement, grief, terror and surprise. The scope for further observation and study in comparative animal and human psychology is enormous, especially in regard to expressive attitudes of a more subtle nature—such as those found in evasion, sympathy, deception, shame and humility.

We are all familiar with the rogue who cannot be demoralised, the person who puts up no defence when caught red-handed at some trickery, who is incapable, it would seem, of sustaining a guilty conscience. Such a person was my old grandmother, who used to cheat at cards. At the end of Grandfather's reproach, even sometimes half-way through it, Grandmother would burst into laughter, and the higher he raised his voice in disapproval, the more she would laugh. With the tears streaming down her face and her enormous bosom shaking with mirth, it became increasingly difficult for Grandfather to maintain his dignity. His case simply crumbled away, and there was little else he could do but join in the fun. This was not taken in good part by other members of my family, especially those who had lost their money. My grandmother's social technique was remarkably like Samba's behaviour in her dealings with us, as witness the following scene.

Liz is asleep in the lap of my sister Lorna, and I can just

see Samba through the connecting door to our kitchen, where she is sitting on the table watching my wife June ironing the baby's napkins. Samba dislikes the iron because it hurts. In monkey logic anything that hurts spells danger, and there are only two ways of dealing with it: either go as far away as possible, or kill it, if you can. Like so many of the things which she has learned to accept in adapting herself to life with a human family, Samba knows that this strange animal, the iron, that hurts but never attacks, must be tolerated. She does not have to be an electrician to realise that nothing is to be gained by prodding, pushing, or turning the iron-animal over in order to see whether the underneath part is more vulnerable to attack. Watching her on this occasion, we see her sitting on the end of the table, her eyes following the movements of the iron, with one foot resting on the end of the ironing board as a compromise. The scene is set for at least a few moments of peace. This is broken by June's voice: "Don't touch those prunes! Now get down and go into the lounge!" I hear the thump of padded feet as Samba jumps from the table, and as I look up a small, black, woolly person comes walking quietly and sedately through the door and passes me on her way over to Lorna. She climbs on to the settee and sits down, a study in tranquil dignity.

A moth flutters against the window above the settee, and Samba recognises in this a delightful and unexpected treat. By a rapid thrust of one hand, the moth is captured; the wings are then pulled off and it is eaten alive. With moth-dust on her nose, she composes herself on the window-sill, chewing rhythmically and contentedly, while her eyes wander round the room in search of the next adventure.

Unlike Liz and Jimmy, Samba is a diplomat and a model of good behaviour. Where the other monkeys get into trouble because of their innocence, Samba avoids it by subterfuge. When Samba is occupied with something she ought not to be doing, one rarely notices it until after it is done.

Returning to our story, Samba is still on the window-sill, and her roving eye has now settled on me. I am not going to allow her to disturb Liz, who is now sitting on my lap, so I give her a severe look and say "No!" This causes her to turn her head abruptly to one side and lift her chin a little in the air, which means: "I dislike you. Go to the devil."

She now goes into what we call her roly-poly act. This consists of describing erratic tiny circular movements in the air with her hands, and at the same time curling her tail up from behind so that she is able to hold and wiggle the end of it over her head. It is a most amusing spectacle, and though we have witnessed it countless times it still reduces us to a state of helpless laughter. It takes on quite a bizarre character when she stands up on two legs and accompanies the act with an eccentric dance, alternating the roly-poly hand movements with tentative grabs at her tail, all the while jigging up and down as though about to make a high dive from the edge of a plank. This ritualistic behaviour is usually a prelude to a high-spirited bipedal jump. It may be a long and venturesome jump, such as from the mantelpiece to the floor, where a slight element of risk adds to the excitement. If it is a short jump, such as the one she is now contemplating from the window-sill to the settee, she will maintain her grip on the end of the tail from the moment of jumping until she lands.

This pantomime obviously expresses the pleasure Samba experiences when she jumps, but I have been unable to connect it satisfactorily with any of the instinctive behaviour characteristics of the woolly monkey. I can, however, hazard a guess: that whenever Samba is excited by the challenge of a long jump, she is also frustrated because there is no preliminary support for her tail. If she were able to grasp with her tail a branch above her head, she could swing downwards and reduce the distance of the fall, or she could hold on with one hand, lowering herself slightly before dropping. The holding of the tail above the head may be a compulsive act compensating for the feeling of insecurity. Perhaps the roly-poly hand movements express the same "joyful frustration"—a wriggling about in space for something that isn't there. If this is the explanation, we still have to account for the fact that the bipedal jump not only is unnatural for a monkey, but invites a greater risk and is more difficult than landing on all fours. Whatever the true explanation, the whole performance is Chaplinesque in its appeal, and entirely Samba's idea. I hope that it will remain in her repertoire for many years to come.

Samba has jumped on to the settee, and is delighted with the roar of applause from her audience. She now makes her way over to the toy cupboard, and on her way walks right over Max, thus avoiding a detour of several feet in either direction. Arriving at the cupboard, she opens the door and sits down with her hands folded and resting on her knees. Gazing vacantly into the cupboard, she contemplates with total lack of interest the playthings before her. Her problem is—how to get to that cabinet way over to the left near the piano, where the sherry is? This requires great skill, perfect timing, and a profound knowledge of human instinctive behaviour patterns. She hobbles silently to the left, pivoting along on her haunches as though heavy with humility, arrives at the cabinet and opens the door with great care.

She is now faced with the most difficult problem of all: lifting a heavy sherry bottle and removing it from the shelf without touching the glasses on either side, and then doing battle with a cork that must be twisted and pulled before it will come out. Since the hand-thumbs of a woolly monkey are atrophied and of no more use than an extra little finger, she may use the opposable thumbs of her feet for a more secure grip. Fortunately it is a small cork with a serrated metal cap, a type that she is able to manage on most occasions with her hands.

It is an eerie yet pleasant sensation to be seated comfortably in an armchair with one's favourite book in the quiet of Kent, to hear the tinkle of glass in a far corner of the room, and to know that over there a little black woolly poltergeist from Brazil is bent on nicking your sherry.

The cork squeaks, but there is no sound as she places it carefully on the floor. The bottle is nearly full and very heavy—infinitely more difficult to manage than a cup or a glass—and it has to be lifted up and tilted slowly. Sherry is strong stuff for the sensitive palate of a woolly monkey, and Samba is not an alcoholic; she wants only the merest nip. With the first taste, she shakes her head violently and smacks her lips with uncertain relish, because the taste both pleases and hurts at the same time. One more nip, and the bottle is placed carefully on the floor in an upright position, and "Operation Sherry" is over. Intervention is now imperative because Liz is about to reach for it. Liz can manage to lift the bottle and drink from it, as she can from a cup or a glass, but she is clumsy and often allows the bottle to topple over when returning it to the ground.

There are now a number of courses open to me. I can show leniency—a gentle reprimand followed by a mutual drink to show that we're still good friends and that life is not based entirely on injustice, even among human beings. Or I can reflect that she has perhaps been getting a little out of hand of late. There was that incident yesterday in the kitchen when she pulled the ribbon out of the typewriter. Tonight, however she is sitting there looking at me in such an appealing way that I decide on method II.

Method II proceeds as follows: I stand up suddenly and dramatically, point my finger in most dominating manner, and in a loud and theatrical voice tell her what a naughty Samba she has been. I walk over and kneel down beside her on the floor, and she makes not the slightest attempt to run away. While I rant and rave, shouting and wagging my finger, a smile creeps into her eyes and the corners of her mouth. Out jumps a quiet "huh, huh", and she leans back on her tail as my threatening finger almost taps her on the nose. She shows her gleaming white teeth, the "huh, huh" increases in tempo and volume, the side-to-side head-shaking becomes more vigorous, until finally she laughs her head off! Samba, like my grandmother refuses to be demoralised.

PART II

*The Animal-Human
Relationship*

8

To Humanise or Not?

Anyone who has kept apes or monkeys will know that they have stronger individual characteristics than dogs. The similarity of the individuals in given species varies in proportion not only to their mental development but also to their emotional level. Hippopotami, for example, resemble each other more than do monkeys, and the mental and emotional responses of the former are unquestionably lower, or if one prefers the word, simpler. Monkeys are mentally and emotionally closer to human beings, and they resemble each other less than do hippopotami. Finally, human beings are less like each other than are the individuals of any other species, and therefore, in this human way of reasoning, they must be placed on a higher level of evolutionary development.

For many people, conclusions of this nature represent a preposterous conceit on the part of the judge. Some people may say: "Are you speaking from the point of view of the hippopotamus, the woolly monkey, or of man?" But if we renounce the *human* standard of measurement in our study of comparative animal behaviour and intelligence, as some philosophical analysts would like to do in their study of the human mind (hence their barren enquiry into the meaning of meaning), there is no standard left by which to measure anything; we are left with the unrelated facts—in the words of a French biologist—that a dog dogs, a rose roses, and a snail snails.

The platypus, for example, is a marsupial that lays eggs and suckles its young, but it is also an extremely specialised animal that represents an evolutionary link between reptiles and mammals, and once we admit evolutionary stages we are on speaking terms with that development and history which must ultimately embrace ourselves. The platypus, no less than any other creature, places its own accent on the unity and development of nature and life, and the fact that it is unconscious of the part it plays in the whole serves only to show that it is not responsible for the whole.

When a primate argues that primates are on the highest level of mental and emotional development, he is of course looking for trouble. But even if man is in many ways the most wretched of the animals on this planet, a little credit must be given to some men for accepting this fact and the responsibility it brings. Though I would prefer the society of my woolly monkey Samba to that of the *homo sapiens* Verwoerd, it does not follow that I support those animal addicts who dote on comparisons expressed in the vaguest terms, such as: "The more I see of humanity, the more I love animals." I never heard one of them say he preferred his goldfish to Bach, his cat to Shakespeare, or his dog to a child. I make this digression with some deliberation, to show that my plea for wild-life sanctuaries, which has already become a target for satirical treatment, is just as much a plea for humanity.

There is nothing audacious in my argument for applying the human standard to animals. The conclusions I have reached are well known to naturalists; but the tendency among modern students of animal life has been to avoid the human approach in the name of scientific objectivity. Trained investigators, or behavioural scientists, as they are now called, are afraid of being charged with anthropomorphism—

of attributing specifically human values and sentiments to animals.

It is interesting to see how important the human approach to animals is for tribal man, for whom a profound zoological knowledge is necessary in his struggle to survive. There are many songs about monkeys and apes, some of which describe their habits with an exactitude that comes from the observations of the hunter. The following song gives the precise names of the trees frequented by the ape, the fruits he eats, a description of his movements, and also expresses the human approach in a most dramatic way:

> Proudly he walks up and down,
> From bough to bough he skips
> On the anag-tree, the ape.
> In his cheek-pouches he sticks
> The sweet fruit of the manow.
> Father, mother, look at him,
> He sees the bateg'n fruit,
> He grinds them in his fists.
> "You, give them to your father!"
> His mother shrieks in warning.

The "savage" does not need to be excused on the grounds that it is natural for him to "humanise" because apes resemble man in so many ways. When the hunter expresses the actions and motives of apes in human terms he adds to his understanding of them.

Monkeys have the reputation of being impossible as house friends, but we would not think of saying the same of our children, though we know that if they were not disciplined they would create as much havoc as any monkey. Monkeys are trouble simply because a whole world of do's and don'ts is open to them that a dog cannot enter. One does not say to a dog: "Stop pulling the baby's hair", "Don't put your hand in the jam-jar", "Stop pulling the strings on that guitar", "Put that cup down carefully". You cannot acquire a monkey, take it home, put it on the floor and ignore it as you can a dog, because the psychological and physical machinery of a monkey is so much more developed and adaptable.

Our monkeys have a centrally heated room of their own that measures sixteen by twelve feet, with ropes, bunks, play-material and an asphalt floor with a gully leading to an outside drain. The entrance door connects with a back zoo-kitchen and our laundry, and the top section of this door (as with most of the doors in our house) consists of plate glass. This provides the monkeys with a view of all kitchen and laundry activities and passing traffic from the front of the house to the back, and it also enables us to see at a glance what is going on inside. There is access from this room through a hatch to the outside enclosure. The hatch is self-closing by means of a swinging door, and this they open themselves whenever they go out or return. The outside enclosure measures thirty by twenty by fourteen feet. It has a natural grass base, climbing poles, ropes, swings, a revolving wheel, and an outside hut.

I have dwelt at length on the importance of spacious enclosures and indoor amenities for monkeys because they are *social* animals. Unless they have their own way of life in a group environment, no amount of humanising as a "domestic pet" will compensate for the loss of a natural life in their own community. The baby woolly monkey Danny (born in our colony) had continual contact with his mother Lulu until he was eighteen months old. He suckled as often and for as long as he wanted to, and he was never very far away from Lulu.

The group watched the actual birth of Danny with astonishment and awe. Climbing skills, play, grooming, sexual behaviour, reproduction, the various ways in which ranking order is established and maintained; the bearing, rearing, handling and care of the infant; submission and dominance behaviour, tolerance, bonds of friendship, the whole fabric in fact of the social and biological behaviour of the infant monkey's own species is indispensable for normal and healthy growth. Health grows where health is.

The social meetings between us and the monkeys represent a "meeting point" that breaks through the barriers of species and territories, enriching the way of life of both. The different social lives of both species are temporarily suspended; they compromise; each meets the other half-way. Enlightenment on the significance of a variety of subtle and intimate characteristics in the social behaviour of woolly monkeys depends on a personal relationship at a social level that no amount of study as an "outside" observer can provide. Such a meeting point does not mean that monkeys should be encouraged to live like humans, or that humans should live like monkeys. It means that the friendship and mutual trust that develops as a result of personal contact on a social plane yields a particular kind of knowledge, one that cannot be acquired through the bars of a cage, nor by field studies in the wild. It belongs to a dimension that cannot be experienced by the laboratory worker who is faced by a row of cages, or by the curator who makes his daily round of the zoo and shakes hands regularly with the orangutan. The Samba who welcomes me when I climb a rope to join her on the tree platform is also a dominant female in the monkey community who has the choice of two husbands. She is equally at home in our kitchen or at the top of a tree. She is also the same monkey who attacked a television camera man when he attached a microphone to my shirt.

Such a "meeting point" between humans and monkeys encourages the monkeys to learn new techniques, act creatively, respond emotionally, and communicate on a higher or more developed psychological and social level. The criterion of that level is the human one. There is no other. Such a criterion is not a subjective fantasy, it is strictly scientific; it has its roots in both the functional and historical aspects of animal behaviour. Clearly, if we were to expel all the prejudices of humanising from our observations and interpretations of animal behaviour, an appraisal of the psychological development of the higher primates would not only be impossible, it would be literally unthinkable.

How can we, and in heaven's name why should we, avoid the tendency to humanise? As I began by saying, if we renounce the human standard of measurement, what standard is left by which to measure anything?

9

Memories and Tears

What I choose to call the human approach not only brings us closer to the mind of the animal, but yields its own special kind of knowledge, knowledge which is not quite so unscientific as we have been led to believe. The proposition, for example, that there is in nature a foreshadowing of the moral life of human society is based not on sentimental hope but on the observed facts of animal behaviour.

One of the most illuminating accounts ever given of the animal world comes from the great naturalist Konrad Lorenz. In his wonderful book *King Solomon's Ring*, he writes: "When in the course of its evolution, a species of animals develops a weapon which may destroy a fellow-member at one blow, then, in order to survive, it must develop along with the weapon a social inhibition to prevent a usage which could endanger the existence of the species ... I think it a truly magnificent thing that one wolf finds himself unable to bite the proffered neck of the other, but still more so that the other relies upon him for this restraint. Mankind can learn a lesson from this ... I at least have extracted from it a new and deeper understanding of the Gospel which hitherto had awakened in me feelings of strong opposition: 'And unto him that smiteth thee on the one cheek offer also the other.' A wolf has enlightened me: not so that your enemy may strike you again do you turn the other cheek toward him but to make him unable to do it.''[9]

Lorenz's observation and interpretation of this aspect of animal behaviour are of tremendous importance, not only because they demonstrate a law of nature—that the submissive gestures and social inhibitions in animal life are part of the self-regulating machinery of evolution—but because they show that these instinctive impulses must be accepted as part of the biological heritage of the human race. In the light of this, the evolution of human morals can be seen as a continuation of the process of nature's machinery for defending the survival of the whole as well as of a particular species.

Lorenz is not suggesting that when the young wolf offers its neck to its adversary we have a situation identical with the human appeal for mercy, or that the older wolf who is inhibited from attacking has adopted the moral position of a man sparing the life of his brother, but he is suggesting that in the behaviour of the wolf we have a demonstration, no matter how elemental, of love as the victor over aggression and hate.

In stressing the importance of the human approach to animals I do not wish to give the impression that understanding will be advanced by sentimentality. But I do insist that some of the most significant facts about nature and ourselves will never come to light if our interest is confined to the zoological and scientific aspects alone. A man who has a purely scientific or theoretical interest in music will not only fail as a musician, he will never find the true meaning of music. He must also, as if it needs to be said, love music.

To study animals with profit we must also love or learn to love them. This is not to say that all animals can be loved as individuals. We can smile at the idea of a man kissing his alligator good-night. But it is not always appropriate to smile—witness a true story that comes to my mind.

I was working at the Melbourne zoo. Susan, the chimpanzee, had been working with her trainer at Wirth's circus for seven years, having begun her career as a circus performer at two. It is very rare to find a nine-year-old chimpanzee working as a circus animal, for chimpanzees are considered unreliable public performers from the age of five or six onwards. At the age of nine, Susan was as big as any man, and far stronger. Good-natured though she was, Wirth's circus could not afford to take any chances, and so, after many years of active service with the almost constant companionship of Otto her keeper, Susan had been pronounced unreliable and virtually sent to prison.

Some animals are happy in zoos, and some are not. Susan alone in a zoo cage seemed scarcely living. Her movements were reduced to a minimum. Apart from taking a little food and crawling into a corner to sleep on a bed of straw, she spent her days sitting close to the bars of her cage, waiting and watching for Otto. She responded to no one. When Wirth's circus was in Melbourne Otto would visit Susan every day, but when the circus moved on she would sometimes have to wait several weeks before she saw him again. I was present at one of these reunions after a long absence, and I have seldom been so moved or saddened as I was on that occasion.

When Otto first appeared, Susan's excitement was terrifying to see. She screamed and whooped, shook the bars of her cage, and jumped about like a mad thing. But when Otto opened the cage door and entered, instead of greeting him Susan retired into a corner and with her back turned towards him sat there in silence. Otto sat down, and began to plead with her in German, softly imploring her to come to him. Several minutes passed before Susan turned her head and looked in his direction. Then she came over to him slowly, her great arms lifted and her lips pouting as she hooted and sobbed, and finally embraced him. Otto played and talked with Susan just as any father would with his own child, and when he left about an hour later I hear Susan's cries from the far end of the zoo long after he had gone. A year later, after I had returned to London, I heard from an Australian friend that Susan had died, presumably of a broken heart.

They say that we cannot have the same feeling about an animal as about a human brother. We may share the experience of enjoying a banana with a monkey or a bunch of carrots with a horse, but we cannot enter the inner world of the instinctive and amoral brute, nor he ours. So they say.

The confusion and error begin, I think, with the fixing of an arbitrary division between unconscious nature and human society, between the amoral and instinctive life of animals and the moral and emotional life of man. The distinction is there, and only a fool would deny it, but many other distinctions of equal importance must be recognised before we can pinpoint in evolution anything so pompous as "the rise of the moral law in man".

A certain professor in a television programme on art endeavoured to show that the essential difference between animals and man was that only the latter was able to communicate and share his emotional experiences. To illustrate his point, he contrasted a hen bird that accepted dummy eggs as a substitute for real ones with an aeronaut going into orbit. We were supposed to feel detached from, though perhaps amused and a little saddened by the hen, but stimulated and thrilled by the stupendous human reality of the aeronaut. But if we were to contrast the meeting of

Susan and Otto in the zoo cage with the rapture enjoyed by the professor over the aeronaut, the verdict could be reversed. As an observer of both events, I at least can sincerely say that I was moved by Susan and Otto, and detached from, though amused and a little saddened by the professor.

I hold, then, that some animals, certainly the higher primates, are capable of individual feelings and possess a certain awareness of their own identity, and that they should therefore be treated as individuals and not simply as representatives of their breed.

It has been said that the loss of an animal friend can be replaced by a substitute, whereas the loss of a human friend leaves a space that remains for ever empty. I am not sure that this is true, for tears are tears; we do not weep in one way for an animal and in another for a human being. The grief we feel at the loss of one of our kind who is dear to us may well cause us to cry out: "This is different!" But is it so *different*? Is it not instead so very much greater?

When Jess, our five-year-old woolly monkey, died in my arms, I could hardly see him through the tears. This happened two years ago. We still have five woolly monkeys, and Samba is quite as wonderful as Jess. But she will never be a substitute for the Jess who poked his tail at you while standing up and looking in the opposite direction, chuckling like a demon because he wanted a game. Only Jess would run up the side of a door, wait at the top for your signal, and then pancake on to the middle of the bed.

Jess always gave me a kiss if I was holding him too tightly and he wanted to escape. He had a special way of coming very close and peering intently at any little job I was doing about the home, so close that his nose would follow the movements of the screwdriver, yet he never touched it without permission. His eyes would move up and down, concentrating now on the work in hand and now on my face. As he lifted his eyes to my face and then down again to the job I was doing, his expression was always unmistakable: "What the hell are you up to?" If I swore and cursed, which I have a habit of doing if the job does not go to plan, he would share the experience and express his excitement with a deep-throated "eelk!" When Jess wanted to eat something from your plate, he always said "please" with his own special squeak, and he never touched the plate unless you were looking the other way.

Jess was loved, and he deserved to be loved. Ours was a relationship based on trust, affection and the fun of living. For any animal as highly developed as Jess, there is no substitute. Gavin Maxwell expresses a similar feeling when he tells us about the loss of his otter Mijbil: "I became, during a year of his constant and violently affectionate companionship, fonder of him than of almost any human being, and to write of him in the past tense makes me feel as desolate as one who has lost an only child . . . and though I now have another otter no whit less friendly and fascinating, there will never be another Mijbil."[10]

Somewhere down the evolutionary scale the case for replaceability must hold good. But who can fix it, and where? All we can say is that when we consider such creatures as the moth or the ant, in which the individual is completely submerged in the species, a personal relationship from a human point of view seems impossible and absurd. This is not to say that these creatures, which belong to nature no less than we do, are unworthy of our admiration, for when we turn our attention from the higher primates with their marked individuality to the mole, the frog, even to the very worms in the earth, love and compassion seem to melt into a simple reverence for nature in the general.

The word that most offends me is "substitute". It serves well the language of the critic who says: "His monkey is a substitute for a previous monkey, or for his failure to establish a satisfactory human relationship." I am also unsympathetic to the attitude of the self-tortured: "I shall never have another dog, for no dog could ever be a substitute for Nell. The loss of her is a sadness I cannot bear to think about."

Images of Jess called up from the past fill me with a great sadness, yet I am conscious that a smile is never very far away. Smiling tears express neither an innocent and un-complicated sorrow, nor morbid self-torture. They are those joyful tears that always come with the memories of those we loved, whether they were animals or humans. We cannot forget what we have truly loved. Because of Jess, we gained not only Samba but a greater understanding of Samba, and it is this, not the simple act of suffering, that begets love. The failure to realise this implies that suffering is the main theme of life, that all comes out of it, when in truth all comes through it.

Which of us would not prefer to be remembered with a smile? It would be a poor homage to a lively ghost if he returned to find that those he loved remembered him only with tears. That at any rate is how I feel about a monkey named Jess, and may he never be forgotten.

10
Creative Ability

The order of primates includes man, apes, monkeys, lemurs and many other less monkey-like mammals such as galagos and pottos. Although the zoological reasons for grouping them together are mainly anatomical, the higher primates also resemble each other in a particular kind of intelligence and behaviour. Since a definition of primates in strict anatomical and physiological terms cannot indicate the nature of this intelligence and behaviour, we must invent a term of our own. I shall call it "human-like", but if the bias is disturbing to the scientific disposition I have no objection to calling it "monkey-like", so long as it is remembered that man, too, is a primate. I personally have no objection to including chimpanzees in the order of *Homo sapiens*, provided the chimpanzees are agreeable.

There are those who argue that the human-like intelligence of apes and monkeys is not characteristic of their behaviour in the wild, and that it is only displayed in captivity in imitation of human ways. Although this view is a fallacy, a word first on the "imitative intelligence" of all primates, including humans. Under human influence most of the higher mammals, not only monkeys and apes, exhibit a variety of skills that we could hardly expect them to develop in their natural environment. One is not likely to find a seal balancing a snowball on its nose in the Antarctic. Indeed one can argue that it is only with the help of human influence that animals can be made to look so stupid and to behave so unintelligently. If breeding lines of apes are going to be established in captivity, let us hope that the animals will not be subjected to some of the worst features of human influence. I am not attracted by the prospect of a proletarian chimpanzee that fetches one's slippers, peels the potatoes and does the washing up, and plays the trombone. If the chimpanzee is to suffer the same fate as the modern show dog, then a few hundred years of domestication should be enough to convert him into a hairy, docile, perfumed sub-moron, who brushes his teeth, uses toilet paper, and accompanies his master on a grouse shoot, himself no mean shot. In short, a chimpanzee that does not creep up behind you and give you a thump on the back when you least expect it is no longer a chimpanzee.

Apes do in fact display a number of human-like characteristics in their natural environment. The South African Eugène Marais writes: "It is not only association with human beings that enables apes to evolve their type of intelligence—in a lesser degree—but in its nature there was no difference."[11]

George Schaller, who studied the gorilla in its own habitat, remarks: "A prominent feature of the intra-group relations was the marked individuality in the behaviour of the animals. An observer soon realises that he is not only studying gorillas as a group or as a specific age and sex class, but also as individuals within that group or class".[12]

In my view, the human-like behaviour and intelligence that we find in apes, baboons and full monkeys is an expression of a relatively high stage in the evolution of individuality in nature as a whole; and this individuality is not the sovereign right of primates, be they monkeys or men, but a characteristic that persists, to a lesser extent, in other animal species.

The marked individuality of the higher primates, of monkeys, apes and man, is inevitably accompanied by a higher creative intelligence. My own objection to the use of such terms as "creative intelligence" and "creative ability" is that they emphasise only the intelligence and the inventive ability of animals, and fail to suggest the nature of their emotional life. Whenever I use these terms, I have in mind the emotional and psychological level of the animal as well as its ability to adapt itself to new conditions and its propensity to occupy itself in pursuits which are not dictated by the needs of survival.

The play-activity of most animals has a biological significance: kittens, for example, in specialised movements for hunting and prey-catching, are developed and rehearsed. But much of the play-activity in monkeys and apes is due to their craving for variety; it is not determined or driven directly by the needs of survival, though it will obviously be influenced by the latter.

Whenever I speak of the instinctive behaviour of an animal as distinct from its creative ability, I mean those innate movements which express, for example, anger, pleasure, fear and submission, gestures of threat, prey-catching movements, the carriage of body and tail, and the many species-preserving functions which are commonly referred to as instinctive behaviour patterns.

It is possible, of course, to consider animals from the point of view of their aesthetic appeal. Few animals compare with the cat for sheer grace of movement. No matter how much we may admire the functional design of the wart-hog, we must agree that he is not exactly a good-looker: magnificent

perhaps, but not beautiful. Whether such aesthetic comparisons are objective, or only express our personal feelings, is
irrelevant to the present question. I shall not consider the
"entertainment value" of instinctive actions, such as the
grooming and feeding habits of the hamster or the storing
mania of the mongoose—rituals that are so appealing and
impressive that the observer is tempted to regard them as
indications of a high intelligence. Nor shall I compare the
instinctive skills of one species with another's, for how could
we compare the instinctive intelligence of the beaver with that
of the spider without finding ourselves in the ridiculous
position of trying to decide whether the beaver displays more
intelligence and imagination in the building of a dam than the
spider does in spinning its web?

My purpose will be to examine the relation between the
behaviour instincts and the creative ability of a given species,
and to compare the behaviour characteristics and creative
intelligence of various animals one with another. Although the
scope for comparative studies of this nature is unlimited, I shall
keep mainly to those animals that are well known to us all—the
cat, the dog, and the monkey.

I once heard a man argue that his cat was more intelligent
and purposeful than his dog, because the cat persisted in
jumping on to the table for food in defiance of repeated threats
and even punishment, whereas the dog, once corrected,
waited obediently until permission to take the food was
given. In this situation the dog's capacity for learning and
profiting by experience is clearly greater than that of the cat,
from which it follows that the dog is more adaptive as well as
more responsive to human influence. It is this very quality in
the dog that enables it to act creatively, not only in such
simple gestures as the giving of the paw, but in its protective

attitude to man and the initiative it displays in conveying its needs and feelings.

Unlike the dog, the cat is not a social animal. Regardless of its domestic history, it has remained a "wild" creature, not in the sense of being aggressive or savage, but in its self-reliance. The modern domesticated cat is generally considered to be a descendant of the African wild cats, the serval and the caracal, and these beautiful animals, unlike the European wild cats, who are virtually untamable, are well known in zoos for their tractableness. The wild cat is not a social animal, and it stalks its prey alone. Its innate drives and responses conform to an instinctive pattern that resists human influence, and although it has become more docile in the course of domestication, its essential independence and remoteness remain almost unchanged.

The character of the dog, like that of its ancestors the wolf and the jackal, has its roots in the pack instincts of leadership, submission, mutual aid and loyalty. The dog, therefore, has a wider range of instinctive actions and responses to draw upon than the cat, and thus a greater capacity for responding and expressing itself creatively in association with man. The fact that the cat has a shorter domestic history than the dog has little bearing on the point, for even if their domestic histories were equal in length, there can be little doubt that their characters would remain distinct. The essential difference rests in the fact that the domestic cat *continues* to be an independent animal, and the dog *continues* to be a social animal. The personalities of both animals remain tied to the innate behaviour characteristics of their ancestors.

"Individuality" in the cat is in fact considerably less marked than in the dog; or, to put it another way, dogs are less like each other within their species than cats. Every animal has its appeal, but for me it is easier to become attached to a dog or a monkey than to a cat. A close relationship with a cat is a one-way affair. You may feel close, but the cat will not.

The dependence of the dog on its master arises out of the instinctive loyalty of the wild dog to the pack-leader. Because of this "transference" in the man-dog relationship, the development of the young dog is sometimes arrested and kept at the infantile stage. Although the extreme "childishness" observable in some dogs must be regarded as a character defect, it must also be seen as the reverse side of a trait that is positive, i.e. a capacity for loyalty and affection of a very high order. In dogs fawning is no indication of healthy obedience, and the extreme pleasure derived from petting cannot be regarded as real affection, yet it would be quite wrong to attribute these traits to an inherent weakness in the character of dogs in general. It cannot be claimed as a comparative virtue that this behaviour is uncharacteristic of the cat, for the very reason that the cat is by nature resistant or incapable of the kind of affection and loyalty that is characteristic of the dog. It is easier to corrupt a sensitive man than an insensitive one, but this does not mean that the latter has a stronger and a larger character, any more than the absence of fear presupposes courage. The mental and emotional life of the house dog, the wolf and the wild dog functions on a more evolved or higher level of instinctive behaviour than that of the cat, which is precisely why the dog is more responsive and creative in his relationship with man. I have never heard of a guard-cat, though I understand that relatively tame and well-fed lions have been known to roam the estates of Eastern potentates.

It is because some animals within the higher mammalian groups exhibit distinct and highly individual characteristics that we call them "personalities". When we study those animals that represent earlier stages in evolution, such as the opossum and the koala, we find one member of the species almost identical in behaviour and appearance with another. If we refer to the koala as a "personality", it is the "personality" of the species that we have in mind, and not the individual koala.

The koala, of course, is an Australian marsupial and not a bear. He may resemble the teddy-bear, but in appearance, behaviour and temperament he is nothing like a bear. As for his cuddlesome appeal, he must be handled with extreme care, for he bites, unpredictably and without warning, and his claws are capable of inflicting a serious wound—as anyone who has been rash enough to climb a eucalyptus tree to catch a koala will know. In the mating season the call-cries of the koala can be heard echoing through the Australian bush. The sound, which resembles the noise of a saw cutting through iron, is as prehistoric as the koala's own anatomy. He is said to wear a smile, but this is due to the shape of his muzzle, for his face, like that of the South American sloth, is permanently innocent of any kind of expression. He is not an emotional animal.

I love the koala as a species, as a manifestation of nature, but communication between me and the individual koala is restricted to one situation. Up a tree in the Australian bush, a picture of vegetative bliss, he looks down at me, languidly and vacantly, while I, at the foot of the tree, look up at him and think how much better the tree looks with him in it—two creatures on the one earth, each gazing shortsightedly at the other, with no enmity whatever, yet neither feeling the slightest desire to shake hands or say hello. All one can do is to give the tree an affectionate pat and wander on to another part of the forest. Nevertheless, it is a wonderful experience to walk through the reserves and see these quaint toy-like creatures clinging to the branches of the eucalyptus tree, serene and detached, as though happy in the knowledge that evolution has passed them by.

If I have been unkind to the individual koala, it has only been with the object of making my point: that as individuality within the species decreases, so does the scope for communication between the animal and the human observer, whether he be a friend, an investigator, or both. That is why even the strongest attachment to an animal can never give expression to the deepest and most urgent needs of man. Yet who is to decide where respect and reverence should end, and love begin? Can the critic who thinks we should confine our "specifically human" feelings—whatever they are supposed to be—to members of our own species, tell us where we should draw the line—at the Tasmanian aborigine, the Neanderthaler, the gorilla, the woolly monkey, the dog?

Just as the dog has a wider range of instinctive social responses to draw upon than the cat, so has the monkey compared with the dog. It is more difficult to assess the monkey's comparative power of understanding human language because he is by nature less obedient than the dog. The monkey that lives under ideal conditions with a human family, such as most dogs enjoy, is frequently preoccupied and gives one the impression that he is not listening or does not understand what is being said to him. We know that the dog is more loyal, more dedicated to his master and the family, that he has a smaller life of his own and that there are all manner of things he can do that a monkey couldn't—I cannot see a monkey competing with a working sheepdog or with a fox terrier when it comes to catching a rat. But if we are to measure animals in terms of their instinctive skills and their usefulness to man, we shall have to bring the cow, the silkworm and a host of other creatures into the picture, and the object of the study will be lost. The special skills which are inseparable from the behaviour instincts of the various species must not be confused with the ability of animals to learn new techniques and occupy themselves creatively in new conditions. If anyone were to ask me why I like monkeys in particular, the best short answer I could give would be: "Because they laugh, sigh, yawn, burp, snort, sneeze, scratch, sniff, squeak, cry, sob, and gurgle, and pick their noses."

The stronger individuality and inventiveness of the monkey compared with the dog becomes apparent when both animals are left to occupy themselves during what, from their point of view, may be described as the dull evening at home. Max is lying on a couch in the lounge, his eyes turned and fixed on me. He can go out into the garden, but this evidently does not appeal to him. He can grunt, sigh, close his eyes or keep them open, go to sleep and dream, or stay awake. We don't really feel sorry for Max. Life could be better for him; but it could be better for all of us. He has the freedom of large front and back gardens that surround the house, he is taken out regularly to the woods and fields near by, and he is often tired out from a day's battle with two infant boys and the woolly monkey Jimmy, all of whom help to make his life more difficult as well as more interesting. So there he lies, resting or waiting, we can never be sure which, a study in expectation. He therefore indulges in the greatest pastime known to dogs: he does nothing.

Now look at Samba. Like Max, she is not feeling very energetic, for she has spent most of the day chasing Liz in the outside enclosure. She has just topped off an evening meal with another apple, but any moment now she will begin to feel peckish. The clock chimes and she frowns as she turns her head, saying "yook, yook" very softly, which means great displeasure. She hates the clock, especially when it chimes, because we will not allow her to open the back of it and investigate the mystery inside. Her eyes rest on a fruit-bowl on the sideboard that contains some nuts. She slides down from my lap, and although on her way over she herself displays no interest whatever in one of the children's toys, her tail does, as if it had said to itself: "We might as well take this along too; you never can tell."

Arriving at the sideboard, the toy still held in her tail, Samba opens a door that swings out, climbs on to it and sits there, looking at the bowl of nuts. Lifting one hand tentatively, she turns her head, and with the hand poised over the bowl pauses for a few moments to see whether I am watching. I look at her, expressionless and silent. She says "eelk", this time a tiny, distant, high-pitched eelk which I have learned to interpret as "All is well, is it, I hope?" I may say: "Yes, Samba, help yourself." If I say nothing, she will anyway. She then comes over to me, climbs up and drops the nuts in my lap, and waits for my approval. If I ignore her, she picks up my hand, the one with no cigarette in it, and places it on the nuts. And the toy, which has remained in the grasp of her prehensile tail all the while, will also be dropped on to my lap along with the nuts. If I now say in a whispering voice; "You've been a very naughty girl," she will lift up her arm and cover her eyes with her hand and sob.

The point I have tried to make here is that so much was

going on about so little, that the journey from my lap to the bowl of nuts and back again was a study in cunning and forethought, resolved in an enchanting mixture of affection and "crocodile tears".

More than any other animal except man, the monkey relies less upon ritualized behaviour and more on communication by different attitudes and expressions of the face. In the evolution of primates the emerging monkey becomes more and more human-like, until, at the stage of "complete" monkey, and not the lemur-like half-monkeys, he is so human-like that many humans cannot bear to look at him, so strongly are they reminded of what they wrongly imagine to be their own shortcomings!

Unlike the hippopotamus, or the dog, monkeys would make good customers for palmists, fingerprint specialists, face-readers and craniologists. With his long, bony and sullen face, his drooping chin and protuding lower lip, Jimmy, like Cassius, has a lean and hungry look. Dogs have marked facial characteristics and expressions, but compared with monkeys they do not show such a variety of individual and distinct physical characteristics. Contrasts more striking and dramatic could of course be made with comparisons of different animals in an aggressive mood, with teeth bared and ready to attack; but if the comparisons are to have any value they must include a far wider range of subtle movements, gestures and facial expressions than the obvious expressions of aggression and appeasement.

Individuality increases where there is a greater variation from type. Next to man, variations from type are greater among monkeys and apes than in any other species of animal. Unfortunately this has led to the view that acquired knowledge and variation from type inevitably gives rise to a loss of natural instincts. It is also argued that man has lost most of his instinctive reactions, and that with the advantages of a word language he does not have to posture like an animal, nor does he have to interpret the eye and facial expressions of other men. So untrue is this that I, for one, give the smallest part of my attention to the words of the newcomer, and all the rest to the stance and movements of his body, the expressions on his face and in his eyes, and the tones and inflexions and colours in his voice; and I pray that he is doing the same with me, because in that way we shall understand and interpret each other faster and more reliably. If this is to be regarded as a more instinctive method of communication, then the more instinctive we become the better. The confusion, as I have said, is the result of regarding instinctive behaviour as the prerogative of nature, and therefore the part of our nature which is least human, a kind of primitive backwater, the part we have inherited from our animal ancestors, the drive of our energies perhaps, but the least human, and of course the least moral from the parson's point of view.

Acquiring new knowledge does not mean that we lose our natural instincts, it means that we transform them into new modes of behaviour and become more instinctively human, or humanly natural. There never was a monkey or a dog that could control its body, face or eye muscles with the skill of a Marcel Marceau or a Chaplin. There are perhaps many men who are indisposed, even insensitive, to the higher (not lower) levels of instinctive human experience and communication. Man's natural ability and responsiveness on this level is a significant part of what is truly representative of human behaviour, otherwise the criterion of human achievement would be some kind of expressionless oratory, linguistic analysis, and the ability to make computers and nuclear weapons. Man is not merely a talking machine. From the human point of view, the instinctive level of man finds its highest expression in the fields of the imagination and in art, but the point I am making is not concerned with art, since no animal is capable of intentional artistic expression. More significant for our present purpose is the fact that the highly complex sensory equipment and muscular control of the human primate enables him to convey his emotions, moods, desires and feelings, instinctively and physically in a way that no other animal can. It is surprising how such an obvious truth has been assailed by those who dote on the "wonderful instinctive machinery" of the non-human animal. Of course animals are wonderful. And man is the most wonderful animal, *as well as the most terrible of them all.*

*   *   *

To summarise: there is a danger of treating instinctive behaviour as though it were simply an obstacle to the learning of new skills and techniques, when in fact it is the ground on which new knowledge and skills are developed. That is why monkeys and apes are more *creatively* intelligent and emotionally expressive than any other animal except man. It seems to me that there is an evolution of individuality at work in nature, and with it a development of creative ability that relies as much on *instincts that are inborn* as it does on the learning of new techniques, a development towards human culture, where individuality finds its highest expression, for better or for worse.

11
Apes and Art

When Samba sits on my lap and peers into my face, she never fails to notice the pimple that sometimes lights up the end of my nose. She shows no interest in the rest of my nose, and gives all her attention to that particular area. Samba has a preference for certain colours, patterns and shapes. An enlarged nose, with a red tip and a dot in the middle, is more interesting for her than a normal one.

The woolly monkey's eye for detail, in close or long distance vision, is astonishing. They will peer and pick at something that one cannot see until it is pointed out with the finger, a tiny hair perhaps on the sleeve of one's jacket. Lying in the grass they may reach out suddenly and capture a minute insect on the wing, a mite so small that only the closest scrutiny of the open palm will reveal it. They detect at once a change in one's appearance, not only something so obvious as a new hat, or a new and odd button on one's house coat, but the smallest pimple or tiny scratch on one's face. This will apply of course only to someone they know well. The tie, button, pimple or scratch must be a new and unexpected addition to the friend with whom they are familiar. If Samba were unexpectedly confronted by a stranger with a hook nose and buck teeth, her response would be quite normal: the nose would be investigated with scrupulous care, and the teeth examined with reverence and awe. But if the stranger were to stay and join our community, his prominent features in the course of time would be accepted as part of his natural equipment. Alarm would only be raised if he should then appear on the scene with a normal nose and teeth.

Bernhard Rensch has shown that birds, as well as monkeys, have a preference for certain colours and symmetrical patterns. His preference tests are valuable because they confirm that an instinct for organisation, symmetry and rhythm is rooted in the very heart of life. Desmond Morris however maintains that apes enjoy aesthetic creation with abstract shapes and that the ape and modern man have much the same reason for painting pictures!

Because apes are our next of kin we would expect to find the instinct for organisation and design stronger in them than in other animals; but only in man is this instinct raised to the power of aesthetic intuition and expression. Konrad Lorenz has indicated that in nature there is a foreshadowing of the moral life of man; but when he refers to the instinctive inhibitions that prevent a wolf from using his weapons against another of his own kind, he does not mean that the wolf is a moral philosopher; the wolf has no moral consciousness, any more than the ape has an aesthetic feeling that *compels* him to create as an *artist*.

A chimpanzee is interested in painting for much the same reason as the human infant. Both love variety, and any new act will be repeated until interest wanes. For example, when an infant first grasps the idea of a circle he will move the pencil round and round, superimposing one circle upon another very rapidly, making several circular movements before stopping. This repetition is driven by the desire for action and movement. The infant is not interested in the result, no matter whether the result is an "aesthetic fact" or a hole torn in the paper. He may in fact look at you and laugh with delight while still drawing the circle. But he will not stop, as though to say: "Behold! I have drawn a circle!" He does not finish a drawing; he stops drawing, or he continues to make more circles, or runs the circles into loops or fans, or obliterates the whole with more scribbling, or stops when the paper is fully covered, or when interest lags, or when he wants his napkin changed.

An animal that can throw sticks, smoke a pipe, catch a ball and return it and point with the finger at new objects, will display a similar initiative and dexterity in anything it does, whether it be drawing with a pencil or playing with Plasticine. We do not regard the chimpanzee as a fundamentalist in the first principles of mathematics because he can count up to seven, and it is equally absurd to present him as an artist capable of pure aesthetic creation with abstract colour patterns because he daubs as intelligently as a child of three. Apes are intelligent, and they have a sense of humour. That makes them wonderful enough. Many years ago at the Melbourne zoo a keeper accidentally left a can of paint in the cage of Tang the orang-utan. When he returned, Tang had painted the rest of the cage, the whole floor, and himself as well. He was covered with paint from head to foot, and it was months before it wore off.

There is no special significance in the fact that chimpanzees are completely absorbed by their painting without the usual rewards of food that are given during the intelligence tests. Many tests of intelligence other than painting require no material reward, for the simple reason that there are two kinds of intelligence tests: those that are interesting for the animal, and those that are not.

Samba will unscrew a lid from a jar, turn a door-handle, pick up a cup and drink from it and put it down again so that it will not tip over, catch a ball, and perform many other little operations that can be regarded as intelligence tests. She is not absorbed by any of these commonplace operations, and she will not open a door unless she wants to enter the room, unscrew a lid unless the jar contains some delicacy, or pick up a cup unless she wants to drink. But if you give her a new toy, or even an old one in which she has a special interest, she becomes absorbed and nothing within reason can distract her attention from it. She often busies herself with a toy telephone receiver—the property of our infant son—which has a moving dial that makes a little bell-like tinkle. Most of her time is spent in trying to take everything apart; the earpiece must be unscrewed, the lead detached, the dial removed. Her concentrated occupation with simple mechanical gadgets is for its own sake; there is no question of a reward. Similarly, if Samba is given a piece of rope that is tied in several knots, she will give herself over entirely to unravelling it. If all this is so of a woolly monkey, whose intelligence and initiative are considerably less than a chimpanzee's, it is only to be expected that chimpanzees will become absorbed in a variety of play activities that bring no material reward, and that painting will be only one among many.

The hand of the higher primates is designed to perform a number of complex rhythmical movements, and in a welter of pencil-scribbling and paint-daubing it is unable to avoid making horizontal, vertical and diagonal lines, loops and curves, just as a chimpanzee sitting at the piano will sometimes play major and minor thirds by accident, and even triads, simply because he has a hand action and fingers that fall naturally on to these harmonic combinations. The chimpanzee interests himself with the mechanical action of the keyboard and with the different sounds he is able to make. It can be said that he is enjoying himself, but it cannot be said that he is creating music, that he is aware of music, or that he is expressing himself musically.

If you give a guitar to a chimpanzee, once the relation between cause and effect is grasped (between the touching of the strings and the sound it makes, as with the moving of the pencil and the marks that appear on paper), a number of interesting things begin to happen. Thumping the strings with the open hand makes a drum-like sound. Dragging the nails of all the fingers swiftly and strongly across the strings near the bridge-piece gives a fair imitation of the Flamenco style of playing. Over a period of time the chimpanzee makes a number of exciting discoveries, and experiments with a variety of attacks on different strings. We are impressed by his inventive intelligence and imitative ability, but we do not regard his spontaneous efforts as basic patterns in "pure music-making"!

There can be no objection to regarding the play-instinct as a form of self-expression. But art is more than self-expression, it is as wrong to look for elements of it in the scribbling of a chimpanzee as to seek music in the singing of birds or the call-cries of monkeys. The birds make delightful sounds, but not *music*.

The belief that apes have artistic potentialities has its counterpart in the Darwinian hypothesis that apes are able to produce and respond to musical tones. Darwin writes: "Primeval man probably first used his voice in producing true musical cadences, that is in singing, as do some of the gibbon-apes at the present day . . . and we may conclude that this power would have been especially exerted during the courtship of the sexes."[13]

In the field studies of trained observers there is no evidence that apes utter a sex-call, least of all a "song of love", during a preliminary stage of their mating. Schaller, in his study of the mountain gorilla, noted a staccato cry, and loud growls and grunts which occurred during copulation, but he heard no cries that could be connected with sexual overtures or "courtship". The initiation of gorilla sexual behaviour is expressed by strutting displays on the part of the male, or by receptive females presenting themselves. I have been able to study a large range of call-cries in our woolly monkey community, but none of these is connected with mating behaviour during initiation or copulation. They are signal calls, expressing "All is well", "Danger", and variants of these. Unlike the copulation sounds, which consist of sighs, sobs, grunts and squeaks, and are anything but musical, their call-cries are among some of the most beautiful sounds I have ever heard in nature.

Darwin's approach to the origins of music is so naïve that it is difficult to take some of his conclusions seriously. Take, for example, the following quotation: "Judging from the hideous ornaments and equally hideous music admired by most savages, it might be argued that their aesthetic faculty was not so highly developed as in birds."[17] Darwin evidently preferred the chirruping of canaries to the beautiful chants and dynamic rhythms of Australian aboriginal music, and believed that apes and early man sang for their mates.

In informal language we may say that gibbons sing and that chimpanzees dance, but we do not mean that gibbons sing music, or that chimpanzees dance to or because of rhythm.

If they dance rhythmically it is because they find it extremely difficult to jump up and down unrhythmically, just as we would find it difficult to breathe irregularly.

I have discussed the question of chimpanzee "dancing" with many circus trainers, and they all agree that the chimpanzee has no real sense of rhythm. This is the opinion of the famous chimpanzee trainer Rudi Lenz. His chimpanzees enjoy themselves in their "jazz-band" act. They are excited and stimulated by the music, and each performer will stamp his feet, beat a drum, and hit a cymbal with wild enthusiasm. But Lenz insists that not one of them, as an individual or in the group, has ever displayed a rhythmical ability remotely connected with keeping time to music.

Almost every specialist in ape behaviour has commented on the alleged "rhythmic dancing" of chimpanzees both in the wild and in captivity. Even trained investigators like Köhler and Yerkes have claimed that chimpanzees are able to dance "in time to music", but they produce no evidence to support these conclusions. My own observations and tests indicate that chimpanzees will exhibit rhythmic action in their intimidation displays, and also when they are frightened or excited. These displays consist of much foot-stamping and "drumming", jumping up and down, rocking from foot to foot and swaying rhythmically. A parrot on its perch will also rock and sway in rhythm. It would be an eccentric bird that did otherwise. The sustained repetition of any action in the animal world would not be possible without a certain amount of regularity, but the rhythm implicit in such regularity is determined by motor impulse and not by a sensuous and rhythmic feeling in the psychic life of the animal.

Though we cannot credit animals with either a musical or rhythmic sense, an instinctive rhythm is nevertheless deeply rooted in the heart of nature; in breathing, walking, the act of procreation, the dripping of water, the flapping of a bird's wings. It has even been suggested that the repeated syllables of infant babble are to some extent influenced by the sounds of the heart-beat first heard by the embryo in the mother's womb.

The rhythmic sounds of nature are impressive and mysterious, and they belong to a different order of sound to the warbling of birds and the cries of monkeys and apes. "There is a sound which is fraught with mystery, a sound which is Nature's magic, for by it dumb things can speak . . . When that strange and curious man first struck together his two pieces of wood, he had other aims than his own delight in sensuous sound; he was trying to create something that had mystified him. That something was rhythmic sound— on which roots the whole art of music."[15]

Apes can be puzzled and intrigued, and they have a sense of fear, but they have no sense of mystery, and they have no sense of rhythm. Man was the first animal to discover rhythm and become aware of the mystery of rhythmic sound. He danced this rhythm to a stage of rhythmic trance in his fear and worship of the unknown, unseen forces in nature. When he added his own speech sounds as well as the emotive cries of his primate ancestors to rhythmic action and chanting in his magic ritual, he was fashioning the true beginnings of music and dancing as art.

In his fertility-ritual and hunting-dances and in his cave paintings, Paleolithic man sought a magic control of the mysterious processes of birth and generation. Like his primate ancestors he was ignorant of these processes but unlike them he was moved and disturbed by the "knowledge" of mystery, by a respect and fear of forces he could not understand.

The ape is aware of what he sees and he knows well where to find his food. He is alert to the dangers of the forest and skilled in avoiding them. He is aware of disease and knows that it brings pain. He is frightened by tropical storms and by strange sounds that come from objects and places that he cannot see. He does not know why the food he gathers must grow, and why it is sometimes scarce, nor why the snake is poisonous, the night cold, or the sun warm. The mother chimpanzee will carry her dead baby around while it decomposes in her arms; she is distressed by the fact that it does not behave as though it were alive, *but she does not know that it is dead.*

Neanderthaler knew, doomed though he was to extinction. He buried the bodies and the bones of his dead and coloured them with red ochre. Red is the colour of life and health, and since shells and ornaments of bone and ivory were also deposited in the graves, it is clear that Neanderthaler had some kind of belief in the supernatural. He survived the Ice Age of the late Pleistocene period. He made flake tools, a flint-tipped spear, hunted the rhinoceros, and taught his skills to his children which required speech and discipline. No doubt Neanderthaler was as much in awe of nature, and danced and sang no less than the men who came after him.

The supreme difference between the ape mind and the mind of man consists in this: man is *aware* of what he cannot explain; he is conscious of mystery, and not simply mystified. He must explain and express the unknown and the unseen, for there reside the powers which control his life. This I believe is the key to the religious, moral, and artistic life of the first men.

12
Conflict Situations*

The behaviour of monkeys and man is remarkably similar. Our woolly monkeys will often laze in the sun on their backs, with legs splayed and arms thrown back above their heads. When they wake up in the mornings they stretch their arms and yawn. They jump up and down when excited. When they are frustrated they shake their heads in anger and sway from side to side. Samba, who happens to be the only monkey in the colony who refuses to come to terms with donkeys, is sometimes inconsolable when distressed by the approach of Prosper at the kitchen window. She will grasp the rail of the

---

*This chapter was written after the publication of the first edition of *Samba* in 1965. I have included it here because it continues and—sadly—completes the Samba story at the Monkey Sanctuary in Cornwall. It was written a few months before Samba died in 1972.

The dialogue between Samba and me is by no means pure fantasy. It expresses what Samba did indeed teach us by her behaviour. Fantasising, humanising, call it what you will, it is nevertheless based on observations and incidents that were objective and real. When the Amazon hunter dramatises the emotive sounds of monkeys in ritual and song, he is not being anthropomorphic, he is empathising with the reality of a forest life he shares with them. Similarly, "Conflict Situations" was written in the climate of an inter-living relationship between a colony of woolly monkeys and a human family, at a time when the birth of babies and the emergence of new monkey characters was making life more exciting as well as complicated for both Samba and me.

chair on which she is standing and shake herself all round the room.

There are many types of social behaviour which man shares with monkeys. A human parallel can be observed in the submissive gesture of a woolly monkey when he crouches down on his forearms and covers his eyes and mouth with his hands, comparable with the submissive behaviour of man when he begs for mercy, and in the greeting ritual where the nodding and lowering of the head expresses friendship. A quick sideways turn of the head or monkey "cut-off" is similar to the human headshake meaning "no". It can also be a gesture of disapproval, or unwillingness to respond. In monkey as well as human language it can mean: "I want no part of this"; or "I am uninterested". It is often accompanied by an expression of disdain, as though the overtures from the adversary were unpalatable, just as the argumentative human will sometimes turn his head away and frown, as though the point made had given him indigestion.

Monkeys and humans have pent-up emotions, and they like to participate in the dramatic behaviour of others. In the early days of our colony a mock battle between Max and myself would cause great excitement in the monkey community. Today the event has become commonplace, and affords no more interest than the conjuring trick that is performed once too often for a human audience. Monkeys however are great spectators, and in the excitement of the event they will release tension, by jumping up and down and even throwing objects about, just as human spectators at a sporting event will chant, clap, stamp and throw their hats into the air.

Conflict situations are not always easy to define. When Jimmy is chased by Max over a long stretch and can sense or hear Max gaining ground, he is by no means terrified, nor does he behave as though he were being chased by a dangerous aggressor. He knows it is a game, but nevertheless the excitement introduces an element of uncertainty bordering fear. Max weighs about one hundred and forty pounds, and to have him bearing down on you on a grassy slope at twenty miles an hour is no joke. In this predicament Jimmy usually stops running and turns round and faces up to Max. He gets swept off his feet, but by facing the dog he is able to leap up and hang on to his jaws, making the encounter as painful for Max as he can.

In the conflict situation known as the critical-flight distance, there is a safety limit beyond which the animal will either flee from danger or attack the intruder. Changes in the behaviour of animals occur at critical stages in the shortening of this distance, and it is important for the intruder to understand them. It is no less important for the predator to know when to attack than it is for his prey to know when to flee. But when two animals of different species that normally have no contact with each other become involved in a conflict situation, the social signals of threat, appeasement or escape peculiar to the different species will often break down, and with results that are sometimes disastrous. The behaviour patterns of a donkey for example are entirely different to those of a woolly monkey, and some strange things begin to happen, as we shall see, when these two animals come face to face for the first time.

With the advantages of a second opinion in mind I have, in fantasy, consulted Samba on the question of conflict situations:

WILLIAMS: Broadly speaking, do you agree with me?

SAMBA: For once, indeed I do. There is far too much talk about animal psychology and not enough talk about *space* behaviour.

WILLIAMS: What do you mean by space behaviour?

SAMBA: Well, my opinion of a donkey varies according to where the donkey is. I think a donkey is at its best when it is completely out of sight. It is at its worst when it comes walking into the kitchen.

WILLIAMS: Isn't that the same as saying you don't like donkeys?

SAMBA: No. You will argue about the things you like or dislike, regardless of where they are and whether they concern you or not.

WILLIAMS: I don't follow.

SAMBA: I'll put it as simply as I can: I don't dislike or like Prosper. I like or dislike *where* he happens to be. I *like* him in the donkey paddock. I *dislike* him in the kitchen.

WILLIAMS: Does your theory of space apply to grapes?

SAMBA: Of course it does. I like the grapes I eat. I don't like the grapes you eat. The grapes in my mouth are differently placed. This is a matter of space. Surely you can follow that?

WILLIAMS: I can see I'm up against the limitations of the monkey mind. For me, grapes are grapes, no matter who eats them. I'm talking about grapes. You're talking about who eats them.

SAMBA: How can anything sensible be said about grapes if you don't eat them?

WILLIAMS: Could we come back to the question of conflict situations?

SAMBA: Certainly. I wasn't aware that we'd left it. I run after grapes when they try to escape in the visitor's lunch basket, and I run away from donkeys when they come too close.

WILLIAMS: But that's just the point—you don't run away from donkeys, you attack them! Only yesterday I dragged you off Prosper's leg.

SAMBA: That happens to be Prosper's approach distance—a bite in the ankle—the critical moment when he takes flight and goes careering round the sanctuary.

WILLIAMS: But I'm not talking about Prosper's flight or approach distance. He hasn't got one. From the point of view of a wild monkey he's a domesticated twit, although a good natured one. I want to know something about the flight distance of the woolly monkey, and I want to know why you attack Prosper, a stallion who weighs nearly four hundred pounds.

SAMBA: I'm a great believer in threat. It works well in our colony, but it doesn't work at all with Prosper. I'll be frank and admit that I don't understand him. I scream, I adopt the most threatening postures known to woolly monkeys, but he goes on chewing grass as though I were not there. This infuriates me.

WILLIAMS: Why don't you run away when you see him coming?

SAMBA: I do, but he often appears as though from nowhere—when the nearest tree is too far away to reach with safety. What else can I do but attack? D'you expect me to "freeze", petrified with fear, like one of the lower order of mammals? I've seen this done before in the Amazon, and sometimes they get eaten just the same. I admit I'm afraid of Prosper. He's too big, and I don't like the look of his feet. I've seen tapirs rushing along the banks of the Orinoco river at a frightening speed. If a girl weren't quick off the mark she'd be trampled to death. But there are plenty of trees in the Amazon, and much better ones than those you have here in Cornwall.

WILLIAMS: How do you account for the fact that you are the only monkey in the colony that objects to Prosper?

SAMBA: With the exception of Jojo, I am not impressed with the general standard of intelligence in our monkey community. Jojo is not afraid of Prosper because he is not afraid of anyone. Not even of you. That's one of the reasons he became leader of the group. As for the others, apparently they are not intelligent enough to be afraid. I can understand juveniles like Jimmy rushing up to Prosper without alarm. They treat him as though he were a large dog, but one of these days they'll discover how

untrue this is. I've seen Prosper kick Max, and
I know better.

WILLIAMS: It was of course silly of me to expect your
behaviour with a donkey to conform with any
of the known ways in which a monkey resolves
a conflict situation in her own society.

SAMBA: What known ways?

WILLIAMS: There is the well-known monkey "cut-off".
You often turn your head quickly to one side
when Lulu approaches or passes you on her
way through the corridor cage. As a subordinate female you have no desire to prrovoke her.
The cut-off is your way of acknowledging her
right of way or place.

SAMBA: This is sometimes true, but not always. The
gesture is also made by a *dominant* monkey to
a *subordinate* one. Jojo is leader of the colony,
but he will often display the cut-off when
Django or Jimmy wish to pass him in the runway. He is simply telling them they may pass
him without harm. There is a danger of over-
simplifying the gesture of cut-off. It does not
always mean: "I want nothing to do with
you." Monkeys, like humans, do not like to be
stared at. Turning one's head politely to one
side as someone passes by is a courtesy. Even
some humans do it. Only rude people stare at
you as you pass by. There are many different
kinds of cut-off. In the defensive cut-off, the
head is not only turned to one side; it is also

lifted sharply in the air, and with a slight lowering of the eyelids. In the courtesy cut-off, the head is turned away, but it is *lowered*, not unlike the bowing of the head in human ritual. Perhaps you are a little confused because woolly monkeys are more polite and sensitive in their social behaviour than most humans?

(Seated on the top of a grass slope which leads down to the monkey territory, two visitors are preparing a picnic lunch. Immediately behind is a mass of shrubs, vines and saplings, into which Samba and I disappeared a few moments before the visitors decided to graze in that area. Shaded from the heat of the afternoon sun, Samba and I are far too comfortable to worry about the ethics of eavesdropping.)

VALERIE: How nice to see monkeys in the trees.

HENRY: One of these days they'll scamper off into another part of Cornwall.

VALERIE: They seem to be happy here. Why should they want to escape?

HENRY: Monkeys are nomadic. They forage for their food. Sooner or later they're bound to explore.

SAMBA: (Tell him that a colony in the Amazon moves to another area only when the food supply runs out. And while you're about it, explain that we do not scamper. We walk, run, climb, leap, swing and brachiate, but we do not scamper.)

HENRY: The explorer Humboldt referred to a tribe of

Amazon Indians that ate three thousand woolly monkeys in a year.

VALERIE: I wonder what they taste like!

SAMBA: (One primate tastes much the same as another. If she really wants to know, why doesn't she eat Henry?)

HENRY: Strange they should sleep in the monkey houses at night instead of the trees.

VALERIE: It must be too cold for them outside.

SAMBA: Do they really expect us to sleep in the crotch of a tree, when there are blankets, sleeping bunks, and midnight snacks in the monkey house? If the sanctuary were linked by ropes to Brazil, every primate in the Amazon would come over here.

WILLIAMS: I'm flattered to hear you say that, but I don't believe it for a moment. I think you've acquired a taste for an artificial way of life. How can you prefer the sanctuary to an idyllic tree life in the Amazon?

SAMBA: I think you're becoming a testy old man. You're too critical of human behaviour, and I'm tired of all that nonsense about an idyllic life in the Amazon. Have you ever met the *isango*?

WILLIAMS: What on earth is that?

SAMBA: A minute blood-sucking insect that infests

vast areas of the jungle from central Brazil across to Ecuador. Idyllic! Compared with the sanctuary, the Amazon is an evolutionary dead-end. I wonder whether they've got any tomatoes in that lunch basket.

VALERIE: I say! Just look at that monkey leaping through the trees. Did you hear that! A branch snapped, and the monkey dropped to the branch below. There *must* be an accident sometimes!

SAMBA: (We've all been trying to break that branch since the first day we went to the trees. I'm pleased Jessy has succeeded at last. Do people really think we go prancing about in the trees without knowing what we're about? I think I'll stroll over and cadge a tomato.)

VALERIE: Look, we've got a visitor!

(Samba sits politely by the lunch basket. Unlike the juvenile monkeys she does not help herself to loot and rush away with it. She squats, her hands folded and her arms resting on her knees, and she waits. She discovered long ago that she is irresistible. She is however reasonably impatient, and if the visitor is deluded for too long in the belief that she has come to see him instead of his lunch, she will toy with the lid of the basket and say "eeolk". There are at least ten inflexions of "eeolk". This particular one means: "I'm very pleased about the prospects of a tomato. I hope you are too—from my point of view." Henry takes a banana from the basket and offers it to her. She pushes it to one side and rummages through the basket for something more interesting.)

HENRY: I've never known a monkey to knock back a banana!

VALERIE: It may fancy something else. Try a cucumber sandwich.

(Samba takes the cucumber sandwich and rejoins me in the shrub.)

WILLIAMS: Any luck?

SAMBA: Yes, but I can't stand white bread, and I object to being called "it". It should be obvious that I'm a "she".

WILLIAMS: You can't expect humans to grace a monkey with the pronoun "she". As a lower primate, my dear girl, you're an animal, and therefore an "it".

SAMBA: How does one become a "she".

WILLIAMS: You've got to have self-identity.

SAMBA: What the devil is that?

WILLIAMS: In philosophical parlance, you must *know* your identity. You must be aware of the fact that you're an "I".

SAMBA: Go on.

WILLIAMS: One of our philosophers, Descartes, put the whole thing in a nutshell. He said: "I think— therefore I am!"

SAMBA: What kind of nutshell?

WILLIAMS: I don't know.

SAMBA: I'm aware of the fact that I want a tomato. I think I'll go back for another look.

(The picnic is over and everything has been packed away except a thermos flask. This is lying on the ground unopened. Samba picks it up, unscrews the lid, and drinks. Henry tries to pull the flask away, but Samba's grip is firm and she screams with anger when he persists.)

VALERIE: That's hot coffee! It'll burn itself!

HENRY: There's little I can do. I don't intend to get bitten.

VALERIE: I'm very annoyed. I was really looking forward to that coffee.

HENRY: You've got to admit the flask was put down carefully. I've never seen a monkey do that before. And there's plenty of coffee left for both of us.

VALERIE: I just don't like the idea of drinking from the same flask. They ought to keep the monkeys away from the visitors.

(Samba returns to the shrub)

WILLIAMS: Did you get your tomato?

SAMBA: No. I had some coffee.

WILLIAMS: You don't usually drink coffee. Was it to your liking?

SAMBA: Too much milk, not enough sugar, and the idiot tried to take the flask back before I'd finished drinking.

WILLIAMS: In human society you're supposed to ask for what you want. Did you?

SAMBA: Of course I did. I said eeolk in the best possible way, a polite high-pitched eeolk, with a falling glissando. But like most of the English they speak only one language.

WILLIAMS: Then what happened?

SAMBA: I put the flask down, which seemed to surprise them. I don't know why you waste your time giving those talks in the forecourt.

WILLIAMS: That's not fair. Most of the visitors do listen, and some of them are tremendously impressed. What about the old lady from Liskeard. She comes all the way here, regularly each week, just to see you. I think you're peeved because Valerie wouldn't drink from the same flask as you.

SAMBA: I can't stand racial prejudice.

WILLIAMS: Nevertheless you should have let Valerie drink first.

SAMBA: I don't like lipstick with my coffee.

WILLIAMS: Then you must learn to adapt. Now where are you going?

SAMBA: I can hear potato crisps. They've opened the basket again.

WILLIAMS: Wait. Here comes Jimmy. He's enough for them to cope with.

(Jimmy grabs the potato crisps from Henry and makes off with them.)

VALERIE: He wasn't as polite as the other monkey!

HENRY: They all look alike to me, with their grey coats and black faces.

(Samba makes another attempt to visit the basket, but I hold her back.)

WILLIAMS: You heard them say how polite you are. Why not leave them with a good impression?

SAMBA: We all look *alike*, indeed! You call that a good impression!

WILLIAMS: You must admit woolly monkeys look more alike than humans.

SAMBA: For humans, perhaps. But not for monkeys. For me, Anglo-Saxons are all white, with no tails and blank faces, and most of them aren't worth a second look.

WILLIAMS: Give humans their due. They are more individualistic, at least in their physical characteristics.

SAMBA: I'll grant you that woolly monkeys do not wear bowler hats, glasses and stiletto shoes, and they don't suffer with varicose veins and middle-age spread. I have yet to see a monkey that was cross-eyed, or one with false teeth, or a big female married to a male half her size and weight. I find that altogether rum. It isn't the variations from type that attract my attention so much as the damn silly clothes they wear, and all the paraphernalia they carry around with them. A human female I saw yesterday stood under a tree and gazed up at me through eyes surrounded with black paint. Half the kitchen crockery hung down from her ears, and the smell of cheap perfume that came wafting up nearly lifted me off the branch.

WILLIAMS: Where are you going now?

SAMBA: There are some grapes in the forecourt.

WILLIAMS: But the forecourt is nearly fifty yeards away. You can't see grapes at that distance!

SAMBA: No, but I can see Susan. She's sitting by the porch near the window of the monkey house, holding one hand behind her back.

WILLIAMS: So?

SAMBA: So she wants Jessy to come out with her baby.

WILLIAMS: All right, she's hiding the grapes. But why?

SAMBA: If they all see the grapes there'll be a stampede.

WILLIAMS: So Susan will wait until the others aren't watching, and then show the grapes to Jessy?

SAMBA: Splendid! You have the makings of a behavioural scientist. Poor by Amazon standards of course, but not bad for a human.

Epilogue

I still think of Samba as a person, though she died seven years ago as I write. I would consider myself sub-human if I ever thought of her otherwise. I never had a better friend, monkey or human. When I sit on Monkey Hill today, near Samba's favourite haunt for finding willow herb and grasshoppers, I often fancy I hear her special "welcome" sound—a tiny quiet beguiling "eeolk"—just behind me in the shrubs. The memory of her incredible beauty, and of her loyalty, fills me with sadness. But new faces bring new smiles, and when I see the mother-monkey Jessy and her baby coming up the grassy slope to greet me, I think of the poet Heine when he said: "In the noble pattern of man He made the interesting monkey".

References in Text

1. *The River Amazon From Its Sources to the Sea,* Paul Fountain, London, Constable & Co., 1914

2. *A Narrative of Travels on the Amazon and Rio Negro,* Alfred Russel Wallace, London, Ward, Lock & Co., 1853

3. *A Journey Across South America,* Paul Marcoy. Translated by E. Rich. London, Blackie & Son, 1873

4. *Recollections of an Ill-Fated Expedition to the Headwaters of the Madeira River in Brazil,* N. B. Craig. London, Lippincott, 1907

5. *The River Amazon From Its Sources to the Sea,* Paul Fountain, London, Constable & Co., 1914

6. *The Naturalist in Nicaragua,* Thomas Belt. London, John Murray, 1874

7. *Mammals of Amazonia,* Elaido da Cruz, Rio de Janeiro, 1945

8. *The Rivers Ran East,* Leonard Clark. London, Hutchinson, 1954

9. *King Solomon's Ring,* Konrad Lorenz. Translated by Marjorie Kerr Wilson. London, Methuen, 1952

10. *Ring of Bright Water,* Gavin Maxwell. London, Longmans, Green, 1960

11. *My Friends the Baboons,* Eugène Marais. London, Methuen, 1939

12. *The Mountain Gorilla,* George Schaller. Chicago and London, University of Chicago Press, 1963

13. *The Descent of Man,* Charles Darwin. London, John Murray, 1871

14. *Ibid.*

15. *A History of Music,* John F. Rowbotham. London, Trübner & Co., 1885-87